Spotification of Popular Culture in the Field of Popular Communication

This edited collection considers various meanings of the "Spotification" of music and other media. Specifically, it replies to the editor's call to address the changes in media cultures and industries accompanying the transition to streaming media and media services. Streaming media services have become part of daily life all over the world, with Spotify, in particular, inheriting and reconfiguring characteristics of older ways of publishing, distributing, and consuming media.

The contributors look to the broader community of music, media, and cultural researchers to spell out some of the implications of the Spotification of music and popular culture. These include changes in personal media consumption and production, educational processes, and the work of media industries. Interdisciplinary scholarship on commercial digital distribution is needed more than ever to illuminate the qualitative changes to production, distribution, and consumption accompanying streaming music and television.

This book represents the latest research and theory on the conversion of mass markets for recorded music to streaming services.

Patrick Burkart is Editor in Chief of *Popular Communication: The International Journal of Media and Culture* (with Christian Christiansen). He is Professor of Communication at Texas A&M University, USA, and author of *Why Hackers Win: Power and Disruption in the Network Society* (University of California Press, 2019, with Tom McCourt), *Pirate Politics: The New Information Policy Contests* (MIT Press, 2014), *Music and Cyberliberties* (Wesleyan University Press, 2010), and *Digital Music Wars: Ownership and Control of the Celestial Jukebox* (Rowman & Littlefield, 2006, with Tom McCourt).

Spotification of Popular Culture in the Field of Popular Communication

Edited by
Patrick Burkart

Routledge
Taylor & Francis Group

LONDON AND NEW YORK

First published 2020
by Routledge
2 Park Square, Milton Park, Abingdon, Oxon, OX14 4RN

and by Routledge
52 Vanderbilt Avenue, New York, NY 10017

Routledge is an imprint of the Taylor & Francis Group, an informa business

British Library Cataloguing-in-Publication Data
A catalogue record for this book is available from the British Library

ISBN13: 978-0-367-48346-3
ISBN13: 978-0-367-51566-9 (pbk)

Typeset in Minion Pro
by codeMantra

Publisher's Note
The publisher accepts responsibility for any inconsistencies that may have arisen during the conversion of this book from journal articles to book chapters, namely the inclusion of journal terminology.

Disclaimer
Every effort has been made to contact copyright holders for their permission to reprint material in this book. The publishers would be grateful to hear from any copyright holder who is not here acknowledged and will undertake to rectify any errors or omissions in future editions of this book.

Contents

Citation Information   *vi*
Notes on Contributors   *viii*

1   Lost in Spotify: folksonomy and wayfinding functions in Spotify's interface
    and companion apps   1
    Amelia Besseny

2   Promises and pitfalls: the two-faced nature of streaming and social media
    platforms for beirut-based independent musicians   18
    Chris Nickell

3   Beyond the black box in music streaming: the impact of recommendation
    systems upon artists   35
    Marcus O'Dair and Andrew Fry

4   Revenue, access, and engagement via the in-house curated Spotify playlist in
    Australia   48
    Benjamin A. Morgan

5   Metrics and decision-making in music streaming   64
    Arnt Maasø and Anja Nylund Hagen

6   Digital music gatekeeping: a study on the impact of Spotify playlists and
    YouTube channels on the Brazilian music industry   78
    *Dani Gurgel, Luli Radfahrer, Alexandre Regattieri Bessa, Daniel Torres
    Guinezi and Daniel Cukier*

7   Organizing music, organizing gender: algorithmic culture and Spotify
    recommendations   100
    Ann Werner

8   What do we do with these CDs? Transitional experiences from physical
    music media purchases to streaming service subscriptions   113
    Waleed Rashidi

Index   *127*

Citation Information

The following chapters were originally published in the *Popular Communication*, volume 18, issue 1 (January 2020). When citing this material, please use the original page numbering for each article, as follows:

Chapter 1
Lost in spotify: folksonomy and wayfinding functions in spotify's interface and companion apps
Amelia Besseny
Popular Communication, volume 18, issue 1 (January 2020) pp. 1–17

Chapter 2
Promises and Pitfalls: The Two-Faced Nature of Streaming and Social Media Platforms for Beirut-Based Independent Musicians
Chris Nickell
Popular Communication, volume 18, issue 1 (January 2020) pp. 48–64

Chapter 3
Beyond the black box in music streaming: the impact of recommendation systems upon artists
Marcus O'Dair and Andrew Fry
Popular Communication, volume 18, issue 1 (January 2020) pp. 65–77

Chapter 4
Revenue, access, and engagement via the in-house curated Spotify playlist in Australia
Benjamin A. Morgan
Popular Communication, volume 18, issue 1 (January 2020) pp. 32–47

Chapter 5
Metrics and decision-making in music streaming
Arnt Maasø and Anja Nylund Hagen
Popular Communication, volume 18, issue 1 (January 2020) pp. 18–31

Chapter 7

Organizing music, organizing gender: algorithmic culture and Spotify recommendations
Ann Werner
Popular Communication, volume 18, issue 1 (January 2020) pp. 78–90

For any permission-related enquiries please visit:
http://www.tandfonline.com/page/help/permissions

Contributors

Alexandre Regattieri Bessa, Mphil, Marketing Manager with international experience in managing digital marketing, consumer behavior, strategic planning, and B2B/B2C demand generation. Master's degree in Digital Communication at USP, Brazil, and Post-Graduation Professor.

Amelia Besseny, School of Creative Industries, University of Newcastle, Australia.

Daniel Cukier, PhD, Laura's father, CTO at PRAVALER, co-founder at Playax, PhD in Computer Science, software geek, Google developer expert in Cloud Computing and Virtual Assistant and amateur musician.

Andrew Fry, Independent Scholar.

Daniel Torres Guinezi, Mphil, Mphil in Communication Studies, Bachelor in Economics and Communications, researcher from Datacracy – ECA-USP, Brazil. Data Science consultant for media companies.

Dani Gurgel, Mphil, Brazilian photographer and musician; producing art through digital as CCO of Da Pá Virada. Mphil and PhD candidate at Datacracy, ECA-USP, Brazil. Professor at Senac University, Brazil.

Anja Nylund Hagen, Department of Musicology, University of Oslo, Norway.

Arnt Maasø, Department of Media and Communication, University of Oslo, Norway.

Benjamin A. Morgan, School of Media and Communication, Digital Ethnography Research Centre, RMIT University, Melbourne, Australia.

Chris Nickell, Department of Music, New York University, USA.

Marcus O'Dair, Associate Dean of Knowledge Exchange and Enterprise, University of the Arts London, UK.

Luli Radfahrer, PhD, Associate Professor, Digital Communications, and Leader of Datacracy research group. School of Communications and Arts, University of São Paulo, Brazil.

Waleed Rashidi, Department of Communications, California State University, Fullerton, USA.

Ann Werner, Department of Gender Studies, Södertörn University, Sweden.

Lost in Spotify: folksonomy and wayfinding functions in Spotify's interface and companion apps

Amelia Besseny ⓘ

ABSTRACT

Music streaming sites are growing rapidly and the novel ways in which site users can organize, explore and present their music are important in music discovery. This paper focuses on the organization and presentation of music through folksonomy in the visual interface of Spotify. The site's interface is evaluated to determine its overall folksonomy-friendliness using a framework based upon an analysis of wayfinding features. Folksonomy is a social tagging strategy that exemplifies the innovation of dynamic web interfaces but is surprisingly scarce in music streaming interfaces. It can be found in sites composed of user-made content, where users categorize music in their own words. How music is organized in streaming interfaces, and whether it reflects a user's own musical vocabulary, impacts upon a user's access to a variety of music, their modes of interaction, and overall power dynamics, in which one path of listening may have more influence than another.

This paper focuses on the organization and presentation of music in the visual interface of Spotify. Rather than evaluating Spotify's user interface (UI) for design, this paper investigates taxonomical and folksonomic ideologies behind standard music discovery features, which impact upon categorization. The UX tool of wireframes are employed deconstruct Spotify's layout (Figures 5 and 7) to understand pathways users have available through wayfinding features and allowing us to observe the ratio of folksonomy-friendly functions available. *Interfaces* are referred to throughout this paper in the same manner as Morris and Power, they are defined as "all that greets a user" at the face of a website or app, including content organization and the navigational (wayfinding) options (Morris & Powers, 2015, p. 110). Organizational systems in music discovery, namely taxonomy and folksonomy, constitute approaches to information categorization and classification. Music discovery is the process by which users find the music they listen to. It is apparent through this research that there are few instances of folksonomy in Spotify. One potential reason for this is due to the complications that folksonomy poses for control. Taxonomy is suited for curatorial services, including the automated recommendations that Spotify provides. Simplicity for the end user often takes precedence over the richness of data, as Hogan indicates, invisible algorithms are limited by a reductionism in presentation of information evident in single column ordering and ranked lists, which are inherently taxonomical (Hogan, 2015).

The term *folksonomy* emerged in the early days of the Web 2.0, when dynamic and interactive websites began to replace static webpages. Coined by Thomas Vander Wal in 2004, the term recognized sites such as Flickr and Del.icio.us, which use social tagging to organize and group material. Folksonomy today still involves text-based hyperlink labels but also includes hashtags. Users create tags to categorize their own content or the preexisting content they encounter online, rather than follow recommendations from administrated menus. Folksonomies are "describing tools" and can be used as personal systems of organization and sense-making as users explore catalogs (Jeorett & Watkinson, 2015). On the other hand, taxonomy allows for musical subgenres to be neatly stacked within genres, which creates useable paths for algorithmic song recommendation. This approach to genre though is becoming increasingly complex as databases continue to grow rapidly. Connections between genres also continue to grow as musicians experiment with hybrid styles. Genre maps are becoming de-centralized and look like networks more than trees, such as The Echo Nest project, "Music Popcorn" (refer to Figure 4).

Whilst folksonomy is appropriate for the descriptive labeling of increased amounts of music data, it does have complications in implementation, particularly regarding control. With much music content shared across commercial streaming services, presentation and discovery features are major selling points of any interface, including Spotify. Such music streaming experiences are branded, with services seeking to distinguish themselves in the saturated market. Morris and Powers describe the deliberate yet subtle administrative control in music streaming interfaces, saying that services aim to give the appeal of an "everflowing" and endless stream of music, whilst stratifying for "different levels of consumers, and different groupings of musical consumption activities" (Morris & Powers, 2015, p. 118).

Folksonomy is a tool for engaging with the musical Long Tail in streaming. At the intersection of science and economics, Wired editor, Chris Anderson encountered the Long Tail. Anderson found that where digital retailers were not restricted by what they could stock (traditional supply and demand of physical stock), having a variety of niche products drew more consumers to the service (Anderson, 2010). A major development of moving to digital music has globally exposed the Long Tail, that is, diversity and things outside of mainstream culture. Anderson, noticed the appearance of the long tail in 2004 prompted by Amazon recommendations which used collaborative and content filtering (to recommend similar items, or items other users have enjoyed). The result is a new economic pattern, as consumers may wander further in there listening with increased availability. Researching the Long Tail, Gaffney and Rafferty (2009) advocate for folksonomy as it allows for exploration of the less popular ends of music catalogs and can account for rapidly growing and changing music genre vocabularies.

Some music apps use tagging to bring new approaches to the music archive, exploiting smart technology and social media to gamify online music experiences. Adjacent to this, online folksonomy has progressed beyond simple hyperlinked text-based tags to include time-based tags (annotations) and geotags (tags linking geographical data) across networked environments. Time tags and geotags let users pinpoint moments or map material geographically. Data is shared across a variety of devices promoting a deeper way to interact with media. In the 2010s, there are now many music discovery and music-based social media apps available. Many are aggregators, filtering trending music from various music streaming and media sites (Band a Day, Next Big Sound, White Label), some offer algorithm-based recommendations (iHeartRadio, Discovr,) and are social-based music sharing sites (Cymbal, 8tracks). Innovative apps take a novel approach to music discovery, for example, Songza curates playlists based on moods, WhoSampled shows who artists have sampled in their music and for a brief year Twitter's *#music* pooled music from personal twitter feeds to create a playlist (Gensler, 2014).

Innovations in the presentation of data (and the increased capacity for it storage) have seen a growth in aggregation and customization services. Infomediary apps explore the data of music streaming platforms in novel ways. Spotify-linked (companion) apps provide interesting insights into user data and create novel ways to explore its large musical database. "Forgotify" lets you listen to neglected songs; "Drinkify" matches a drink with your music; "Serendipity" shows the regions of two listeners who are listening to the same song simultaneously; "Climatune", a collaboration with AccuWeather, links weather, location, and mood in playlists; The Echo Nest provides a range of equally exciting apps, including the map generating "Music Popcorn" and "The Wreckomender" (providing antithetical playlists from a chosen song). Whilst companion apps use data to provide insightful curations of music or new ways to manually explore catalogs, the first of port of call is the streaming interface which greets a user with familiar wayfinding features.

Common wayfinding features

Wayfinding features in music streaming interfaces help users navigate ever-immense *streamable* libraries of music. Sites such as Spotify use a range of recognizable features, including menus, headers, featured content sections and customizable search bars and filters for navigation cues. Design in music streaming is faced with the particularly daunting task of making the abstract and intangible (music) into something visual and digestible. Around the same time as the release of the iPod, Maeda wrote *The Laws of Simplicity*, negotiating complexity and simplicity in emerging digital design: "establishing a feeling of simplicity in design requires making complexity consciously available in some explicit form" (Maeda, 2006, p. 45). The field of Information Architecture (IA) grew out of the problem of designing for abundant data, addressing labels and *attributes*, that is, the descriptive details of content and whether the nomenclature (how the labels are named) involves folksonomy. UX design focuses on the user's overall experience of a product or interface. In extension of IA, UX design developed around a value system based upon usability and usefulness, centered on the user's emotional response to interactive services and products (Six, 2014). Alongside UX, UI designers work with the visual elements of the interface, including layout and color, to help interfaces achieve these design principles.

One of the primary concerns of human-computer interaction has been usability, but this has grown to also include where design can promote enriching experiences (Harper, Rodden, Rogers, & Sellen, 2008). Human goals are important to understand for experience design, as Kymäläinen says, because users reveal value in experiences of sharing, creating, collaborating, and configuring (Kymäläinen, 2015). Rosenfeld and Morville wrote the iconic "Polar Bear Book", defining an *information ecology*. The information ecology is a system of three interdependent areas: *(1) Content* refers to what a service hosts, be it images, music, video, or text; *(2) Context*, comprised of goals and resources; and *(3) Users*, the target audience (Morville & Rosenfeld, 2006). Due to the visual format of music streaming sites, labels and written information needs to be succinct and relatable: "the goal of a label is to communicate information efficiently; that is, without taking up too much of a page's vertical space or a user's cognitive space" (Morville & Rosenfeld, 2006, p. 82). Poor labeling can destroy a user's confidence in the site and it can expose businesses that don't have their users in mind.

Labeling systems can be planned and unplanned. Folksonomy falls into the unplanned side, whereas taxonomy is ordered and hierarchical. Drop-down menus and scrollable featured content are planned and presented in a linear order. Streaming interfaces use lines to depict logical, hierarchical and purposeful paths. Straight lines in keep with a longstanding notion of rationality (Ingold, 2016). In planned labeling systems, users only need to learn the hierarchy of system rather than the individual labels. On the other hand, *folksonomous* tag clouds are visualized as clusters, indicating heterogeneity and a lack of hierarchy between content items. Tag clouds may appear irrational but can hold a wealth of cultural commentary.

In a stable taxonomical structure, users are rewarded with easy navigation, but limitations occur over vocabulary and a subsequent bias of viewpoints presented. Folksonomy can potentially help update labels and attributes in IA. For instance, the term electronic dance music (EDM), didn't exist 30 years ago. Pioneered by internet users, EDM and its many related terms (dubstep, house, trap and so forth) have emerged as part of the contemporary music lexicon. Folksonomy-born vocabularies of music genres can come to influence the stylistic leanings of a music streaming service. Such terms have been integrated into organized systems (taxonomies) in SoundCloud, where such genres occur in drop-down menus.

Folksonomy-friendly wayfinding features

Labeling in music streaming walks a line between innovative and familiar. Music streaming has issues that make it abstract and difficult in terms of information design. Firstly, music in audio-only format can be more time consuming to navigate than visual material, which may be instantly recognizable or indicative to users. Secondly, cloud-based music streaming libraries have immense and expandable storage and can support millions of tracks, giving users a surplus of material to sift through. Such factors mean that music streaming platforms rely upon visual and text-based cues for music discovery. Due to this, music streaming sites use navigational cues typical to other web design, but also require some additional features specific to music delivery. Wayfinding functions use a hybrid language, as Manovich says, sitting "between an immersive environment and a set of controls; between standardization and originality" (Manovich, 2001, p. 96). Web interfaces maintain a balance between consistency and originality, upholding common elements "with standard semantics", including "home," "forward", and "backward icons", but may appear different across products (Manovich, 2001, p. 96). Music streaming requires a range of both obvious and innovative cues to accommodate for different approaches to music discovery, that demand curation and self-managed customization. Wayfinding features common across music streaming platforms include tag clouds; headers; drop-down menus; search filters; (scrollable) featured content; queues and player bars.

We can draw upon the discourse of UX design to understand the ways in which users conduct music discovery based upon the wayfinding setup of interfaces. A UX designer's role is to think about the variables that come into interface interaction, for instance, how the infrastructure of a webpage, with buttons, menus, and scrollable functions, will work for a user. Music streaming users may be listeners, musicians or both; UX will aim to consider the varied userbase through hosting, browsing, and social functions. UI on the other hand designs the overall look and presentation of a webpage's infrastructure. UX designers produce prototypes in the form of wireframes, storyboards, sitemaps, and interactive mock-ups, to show the journey of users and to consider the logic of information architecture. Cairo (2012) suggests that we think of visualization as a technology, because in the McLuhanesque sense "it is an extension of ourselves" and "it is a way to reach goals". Wireframes help to elucidate structure in a reductionist model, often a pen and paper sketch of the interface layout. Usually wireframes and flowcharts come in the planning stage of a web interface, before it is made live for consumers, or as part of a design remodel. Later in this paper, the UX tool of wireframes are used to deconstruct Spotify's layout (Figures 5 and 7) to understand pathways users have available through wayfinding features. The precise functions of the figures given are to demonstrate the ratio of folksonomy-friendly functions in current music streaming interfaces. At times, wayfinding functions will allow users to directly add their own input. Such functions are more conducive to folksonomy.

Common wayfinding functions found in music streaming interfaces can be evaluated for a conduciveness to folksonomy:

Tag clouds

Though not the most common feature in music streaming interfaces, tag clouds make user-made metadata visible. Tag clouds offer a spatial, non-linear way to explore categories. They are highly integrated with folksonomy because they stand independent of other hierarchical interface features such as fixed menus. The content of tag clouds changes directly with the input of user data. Whilst some tag clouds may display hierarchy through different font sizes, many use text of the same size.

Headers

Headers break up material visually through the size and style of font. They can be hyperlinks to material or may be plain text. Headers maintain content hierarchy and are usually maintained by site administrators rather than users. Some host platforms such as Bandcamp, Mixcloud, and SoundCloud allow users to create the headings on their own content. Across all streaming sites observed, where users can create playlists they are able to title their own playlists and at times add a descriptive subtitle.

Drop-down menus, side-bar menus

Like Headers, drop-down menus instill a sense of hierarchy into information on music streaming interfaces. Menus are generally made by an administrator using a structured vocabulary. Main menu items don't change regularly but may be updated considering vocabulary shifts. Due to their fixed-ness, menus don't facilitate folksonomy but may lead to a section of the site that does.

Text-box search

Text-box searches allow users to input their own terms as search queries. The degree of auto-finishing or natural language processing will determine how well obscure or misspelt terms will be interpreted to retrieve results. Though the algorithms path is based upon linear logic, this path is not visible to users. The concealment doesn't allow for a user to interact in the way that tag clouds do. Search results may or may not include user-made content, such as tags and playlists. Text-box searches are relevant to folksonomy through user-made queries and may allow users to interact with tags and user-made keywords, though the interaction is mediated through search filters.

Search filters

Search filters can narrow or expand text-box searches automatically or by Boolean phrases. Combined terms (tropical upbeat house music) can refine a search. A search filter can alternatively broaden the search by adding fields to search simultaneously (*tropical, upbeat* and *house music*). Search filters can be useful tools for exploring music folksonomies when they allow searches into tags and playlist titles.

(Scrollable) featured content

Featured content may be sponsored or drawn from users of the site. Featured content is the top tier of visibility in the interface home screen and is presented in visually eye-catching ways. If only paid-curated content appears in these sections, user-created content may suffer in findability. Featured content may be selected by site moderators or may be auto-generated, aggregating trending material. Users' data may become involved in the featured content, but they most likely do not participate by programming it.

Queue

Queue buttons allow users to line up which tracks the player will go to next. Queued songs may be displayed vertically or horizontally, although they may not be displayed at all. Queues allow users to create and set the course of their listening and to temporarily augment existing track listings. Most streaming sites allow for tracks, albums or entire playlists to be queued. In queuing, users may trial combinations of music together. The function may encourage folks-onomy, or result from explored folksonomies, or may not be relevant to folksonomy at all. In this way, queues are useful navigation tools but could be considered folksonomy-neutral.

Player bar

Player bars usually transfer across pages to maintain streaming whilst users browse, search, and queue music. Player bars include common features to navigate playing audio: play button, skip (back/forward), stop/pause, and volume. Some player bars include queue buttons in the section, as well as loop (repeat), or random play functions. Player bars, like queue functions which may be embedded, don't directly impact folksonomy but are an integral part of music streaming interfaces.

Considering these wayfinding functions in the context of music streaming, a scale can be interpreted of *folksonomy-friendliness* based upon hierarchical tendencies and the likelihood of the feature to contain, encourage, or directly influence the entry of user-made metadata (Figure 1). Tag clouds are the most-folksonomy-friendly features, showcasing users' own terms with little moderation and allowing for users to browse other user-made data. Other wayfinding features could be considered more folksonomy-neutral. The contribution of these features to folksonomy depends on the way users employ them, the cues of the interfaces, and where the features are offered in the interface (be it on the home page or subsequent screens).

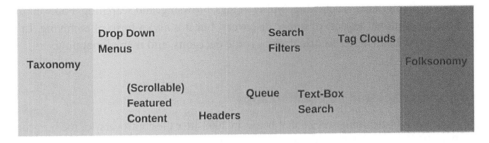

Figure 1. Spectrum of common wayfinding functions in music streaming sites.

The combination of wayfinding features within music streaming interfaces contributes toward how the site will be used and what kind of context it has for folksonomy. Therefore, folksonomy-friendly sites are likely to focus on features that encourage users to search their own queries and generate their own tags. Sites on the other side of the spectrum will use their interfaces to guide users through more complete pre-made paths using featured content and highly-regulated vocabularies in menus.

Each streaming site can be expected to use a mix of wayfinding functions, but the balance and layout varies. Additionally, the nuances of folksonomy promoted in the streaming interfaces can be accounted for using Vander Wal's Broad and Narrow folksonomies. A broad folksonomy benefits the whole user base, as many users participate in it, whereas a narrow folksonomy, no less useful, is created by a singular user or close network, and will be quick to emerge (Vander Wal, 2005; Governor, Nickull, & Hinchcliffe, 2009, p. 62). Broad and Narrow folksonomies may be viewable in the content of the interfaces, but will most likely signal structural features, such as the implementation of wayfinding features, that influence how users employ folksonomy. Logically, if the tags are portrayed on a site visibly, in tag clouds for instance, the folksonomy is more likely to be broad as many users will be able to see the tags and take part in them.

Navigating on the web allows users to move in any direction and to take leaps into disparate spaces. Whilst this affords a kind of choice and freedom, it makes wayfinding complex and vital. It is more than just signage; wayfinding interconnects architecture, landmarks, and interaction (Khan & Kolay, 2017, p. 174). Wayfinding functions are navigational cues to guide users, connect and order information, and allow for searching and browsing activity; therefore, allowing for users to explore the breadth of the site. Lynch and Horton describe wayfinding as made up of four core factors: (1) orientation, locating where you are; (2) route decisions, deciding where to go and how to get there; (3) mental mapping, having a clear concept of where you've been and where to go next; and (4) closure, recognizing you've reached your destination (Lynch & Horton, 1994). Digital wayfinding can be more abstract than wayfinding in the physical world. Describing travel on the web, Lynch and Horton say that whilst being "magical" it "often doesn't provide the concrete spatial and navigational clues we take for granted in the real world of walking through a town" (Lynch & Horton, 1994, p. 1).

Wayfinding plays a part in the general usability of websites and apps but it can also influence the more specific ways that users interact, including the usage of folksonomy. If the site is too geared toward folksonomy, it will be difficult for users to find things. If the music streaming interface is too taxonomic, the music-discovery experience may become a set of generic options and less discovery-based. The listener engaged in music folksonomy is more than just a passive consumer; they are navigating their music streaming course. Users are interacting with interfaces via clicking, typing in tags, browsing, and searching. Folksonomy is part of the navigational aspects of music discovery, but it is also a form of archiving. In this way, folksonomy contributes to orientation, route decisions, and mental mapping.

Spotify's interface and companion apps

Spotify is a Swedish streaming service that began in 2008 after the controversy of the burgeoning Pirate Bay (Wallenberg, 2017). The site began as a true peer-to-peer platform, sourcing music from other servers that owned the mp3s, but has since become its own entity. The site has free and premium tiers, specializing in on-demand service and music recommendation. Approximately 60 million paid subscribers are associated with Spotify (Plaugic, 2017).

Spotify has come up with a range of novel ways to expand users' listening, for example, "Release Radar"[1] and, more recently, "Fresh Finds", a hybrid algorithmic-folksonomic playlist offers personalized curation. Beginning in 2016, Spotify aimed to balance its personalized auto-mated playlist recommendations ("Discover Weekly") with a weekly offering of playlists compiled from early-adopter users ("Fresh Finds"). Adam Pasick wrote of Spotify's interactions with the tastemaking of users, "Spotify is using 50,000 anonymous hipsters to find your next favourite song" (Pasick, 2016). The early adopter is a user that actively seeks out and listens to new music regularly. Spotify's "Fresh Finds" algorithm mines 50,000 of these users for their taste in new material. This is done anonymously without the knowledge of the users (if you are active on Spotify, you may be one of their tastemakers). The "Fresh Finds" playlist is an emergent brand of hybrid algorithmic and folksonomic approaches to music discovery based upon principles of collaborative filtering.[2] The approach is aimed to remedy recurrent user listening behavior, subverting consistent returns to familiar music, which limits the streaming experience.

Despite all the music discovery features that seek to make songs findable on Spotify, there are approximately *four million songs* that have never been listened to.[3] From this daunting figure, auxiliary site "Forgotify" was created by Lane Jordan as an alternative music-discovery tool focused on underappreciated, unplayed Spotify tracks. With esthetics similarities to Spotify's interface in a black, green, and white color scheme, the site plays on themes of the uncharted, with a night sky wilderness background (Figure 2). Forgotify

Figure 2. Forgotify.com.

aggregates *unlistened* tracks by communicating with Spotify's API; the tracks found are played on shuffle through Spotify's embedded player on the Forgotify site, accessed by the "Start Listening" button.[4] The site specifically targets gaps where wayfinding and user-based broadcast has failed.

Users of Forgotify see themselves as partaking in an online musical archeology. It is purposeful discovery, to uncover tracks that the user has not listened to before (let alone any user), or by artists they are unfamiliar with. The app's Twitter feed is full of user discoveries and comments on the obscurity of the tracks (Figure 3). One user remarks that this must be where director Quentin Tarantino goes to find the obscure and kitsch 1970s songs used in his soundtracks.

Forgotify is not necessarily folksonomic, but it is part of a new wave of apps that stem from the same data issues relating to excess. Forgotify and folksonomy alike are a means to reach the neglected ends of music databases. Many sites have an API link so developers can create new apps that explore the data in innovative ways, partly in response to the discovery problem. In 2014 Spotify bought Lamere's The Echo Nest. Daniel Ek, Spotify's chief executive, said of the purchase, "Music isn't just about providing a large catalogue of songs, but about understanding the context in which people listen to music" (Van Leeuwarden, 2018). Lamere and Glenn MacDonald map some interesting correlational music data. Their music analytics about music discovery find interesting assumptions about music genre and draw attention to unusual listening contexts. One such site was

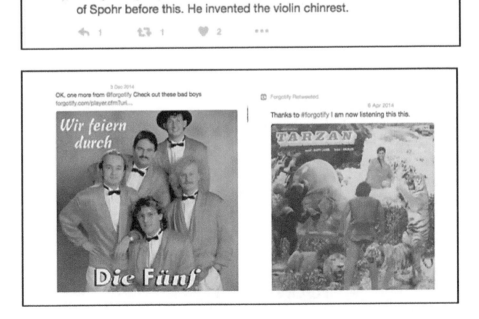

Figure 3. Comments on discovery from Forgotify's Twitter feed.

"Music Popcorn" on The Echo Nest, built by Lamere, which on entering has a scattering of colored dots each carrying a musical genre.[5] The dots bubble around until they rest in a pattern of linked genres (Figure 4).

The electronic music area, which was peach-colored, closely linked to rock and alternative (which was orange). Within the EDM sections there was *grime, trance, glitch, post-disco, new rave* and an array of diverse genres of varying degrees of obscurity. Clicking on an individual *popcorn* rearranges the tags to directly related genres. We see a range of folksonomy-born music genres and micro-genres. Consider, for example, *vaporwave*, a category of Muzak-inspired, pastel visuals, lo-fi glitch music, with an allegiance to the use of the steam-based alternative to cigarettes known as *vaping*.

Despite the sharing of these genre tags across social media and streaming platforms, music is not always classified standardly. In "Music Popcorn", vaporwave was not linked to lounge music like it is on the Wikipedia page (Wikipedia itself is a folksonomy).[6] Instead it was linked to *indietronica* (a mix of indie and electronica), *chillwave, nu gaze* (a shoe-gaze spin-off), *glitch* (evident by the use of broken or malfunctioning electronic instruments), *tribal house, witch house* (house music with an occult esthetic), *footwork* (possibly related to *shoegaze, ambient IDM* (introspective ambient electronic), *wonky* (see glitch, with a pitch shift), *gravewave* (another occult esthetic), *shimmer psych* (related to dreamwave, using ambient synth-pads) and *trap-step* (a fusion of trap and dubstep).

Due to the sheer excess, it seems an impossible human task to map every music-genre tag. SoundCloud and Bandcamp have tag pools based in the millions, and these are frequently added to. Despite the ambitiousness of the task, Macdonald does try to gauge every genre, having created another site linked to "Popcorn Music", the immense "Every Noise at Once".[7] The site, similarly to "Forgotify", interacts with Spotify's API to collate music genre into "a sortable list".[8] Not surprisingly *pop* is Spotify's top genre. *EDM* currently sits at number 12, with *trap*

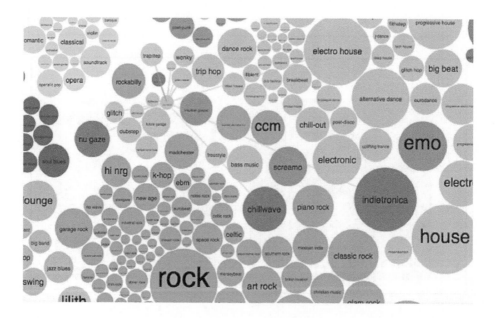

Figure 4. "Music popcorn": Vaporwave.

surpassing it at sixth place, and *dance pop* at third place. Each tag links to an embedded Spotify playlist of that genre.

The immensity of music genre demonstrates the complications related to excess that commercial streaming-services face in information design. When considering the interface itself of Spotify, we can see a mix of features (Figure 5).[9] Considering the checklist of wayfinding functions, Spotify has six of the eight wayfinding features. (Table 1). The two missing functions are tag clouds and search filters.

Spotify has no visible tag clouds, opting instead for menus and scrollable featured content. The text-box search function does allow users to search metadata and potentially to contribute to it, if the site is monitoring user search queries for navigation improvement. New genres are taken on in this way through playlist names, user inquiries and song metadata, keeping up with the more user-generated sites such as SoundCloud. The limitations of the Spotify search box are that users must have a good idea of what they are searching for; they must know the artist's or song's name and at the very least be very close to the correct spelling. Users can manually apply filters using Boolean properties, for example, "artist: Leonard Cohen" for more relevant searching.

Spotify's mobile and desktop apps encourage users to create playlists. The prompts in the interfaces are further supported by the fact that the site does not keep a listening history log, so a playlist may be the best way to return to a song you like. Playlists are also a primary means for users to personalize their music streaming experience in Spotify. As Hagen says, "the playlist is as unique as the user behind it"

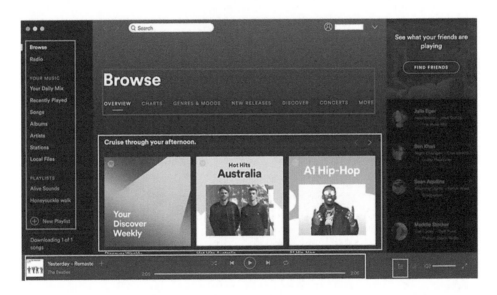

Figure 5. Spotify's wayfinding features in desktop dashboard.

Table 1. Wayfinding features in Spotify.

Present Functions	Absent Functions
Headers	Tag Clouds
Text-Box Search	Search Filters
(Scrollable) Featured Content	
Player Bar	
Queue	
Drop-down/side-bar style menu	

(Hagen, 2015, p. 643). Within playlists, rich user-data can be collected and interpreted to gain insights into intention and identity. Jana Jakovljevic from Spotify says, "playlists not only offer unique insights into a user's habits, hobbies and interests, but provide unrivalled contextual targeting [for advertisers]" (The Drum, 2016).

The unique insights of playlists can also be used in reverse, made by online content creators as branded content or advertising. *Powell Books* in Portland, Oregon, has a series of playlists created by the author to encapsulate themes and moods of their books. Author Rebecca Solnit curated a Spotify playlist to reflect themes of her book *Hope in the Dark*.[10] The playlist consisted of 15 songs centering around notions of empowerment, including the likes of Beyonce, Iggy Pop and Lee Williams (Figure 6). Other examples include the Spotify app, Playlist Potluck, that lets users communally create a playlist as part of a social media event invitation (Section 3.6).

Spotify may have emerged from the rampant music piracy of the early 2000s in Sweden, but now finds itself more formally aligned with music industry ideals (Eriksson, Fleischer, Johansson, Snickars, & Vonderau, 2019).[11] Importantly, the service allocates a large percentage of interface home page space to site-curated or sponsored material. Musicians need a distributor to have music on Spotify, no direct user uploads can occur, but users can curate material in playlists. This may be set to change in 2019 or sooner, as the site trials direct uploads (Kirn, 2018). Popular Spotify playlists are becoming high-value real estate for artists

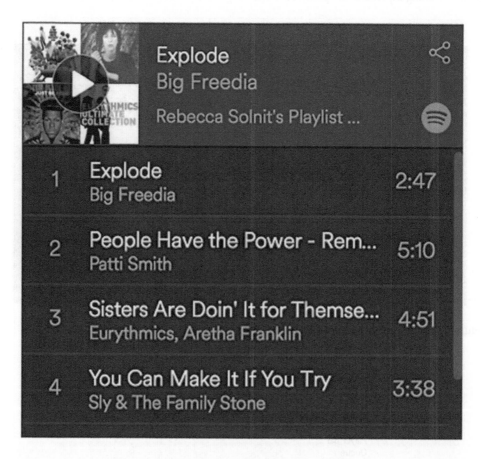

Figure 6. Powell books author playlists: Rebecca Solnit's hope in the dark.

wanting to disseminate their music widely. Playlists and charts are beginning to blur as industry standards such as the Billboard 100 enter the playing field. In 2016, Spotify was suspected of commissioning artists at a lower rate and featuring tracks in influential ambient playlists that received millions of streams (Ingham, 2016). Ambient genre playlists have been particularly successful in Spotify, used widely in the background of office spaces and reception areas, as well as by studying students.

Complications arise in curation between administrator and user because Spotify has garnered a reputation as an aggregator and a tastemaker. Cherie Hu explained this macro lens in the article, *What Is "Escape Room" And Why Is It One of My Top Genres on Spotify?*

> Spotify is a hybrid between a tastemaker and an aggregator. It aggregates our noisy listening activity in real time, then merges man and machine to present this noise through user-friendly channels (playlists like Discover Weekly, Release Radar, Fresh Finds, Rap Caviar, etc.) that ultimately influence our future tastes. (Hu, 2016)

Playlists are Spotify's main area for user-created and user-curated content. They can be shared through linked social media and players can be embedded into external sites. In Spotify's playlists, we can see the presence of non-taxonomical categories, in an interface that otherwise has a strong taxonomical influence as seen in its wayfinding functions, lacking tag-based features. Despite not having many ways for users to practice folksonomy, infomediaries such as The Echo Nest and "Forgotify" find interesting connections in Spotify's data that are visible from the platform's API, but not in the face of the app itself.

Functions for user input only make a small part of Spotify's interface. Users can feed the site data through their playlists and search queries. Nonetheless, Spotify's interface is predominantly composed of curated content through guest curators and algorithmic recommendations. Although users may make tag-based queries in the search box, tags are not shown on the site pages for users to engage with manually or to edit. Other than clicking on items, users can enter text to the site using the text-box search and they can name playlists. Figure 7, in line with UX, is a wireframe, a skeletal image adapted to show only folksonomy-friendly or neutral features in Spotify's Discover page. Text-box search

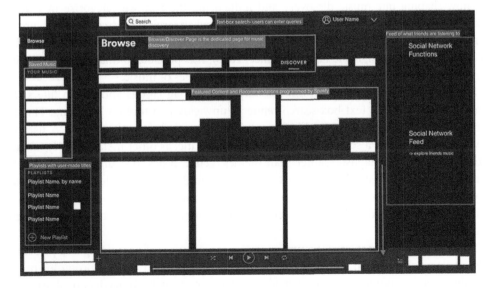

Figure 7. Spotify's discover page.

function, the playlists, a social media feed and the queue (indicated on Figure 7) allow users to add their own keywords or queries. Note that some headers were left to indicate what part of the interface the page comes from.

Conclusions

Music streaming sites utilize folksonomy in their music discovery and archiving functions, such as through tag clouds that users can browse and add to. The most essential features of Spotify's interface are Headers, Player Bars, and Queues, all of which are not necessarily folksonomous tools. Spotify's main folksonomy-friendly feature is the user-creation and naming of playlists. Playlists are searchable and have become important real estate for artists looking to increase their visibility amongst Spotify's vast music library. Whilst the user-made playlist is flourishing, it is constantly negotiated with professional and auto-mated curations, including *Discover Weekly* and *Fresh Finds*.

Apps such as "Forgotify", "Music Popcorn", and "Every Noise at Once", are entryways into the more obscure corners of the Spotify's database and employ folksonomy-like spatial (rather than typically linear) organization. A mutualistic dialogue between AI and human categoriza-tion is beginning to emerge, through the likes of The Echo Nest and Spotify's automated playlists. Elements of folksonomy are becoming integrated into algorithmic processes, such as collaborative filtering, to create more relevant and exciting personalization services. And in turn, users may be able to use automated recommendations to expand their music folkso-nomies. Whilst users can choose much they contribute to the site and whether they actively engage in making playlist names or searching keywords, the actions of their listening sessions strongly influence the direction of their recommendations.

"Forgotify" uses an algorithm to aggregate songs, to identify those that have *flown under the radar*. The creation of this site as an offshoot to Spotify suggests that there were facets of the music library that could not be reached with the current music-discovery functions. The aggregation algorithm of "Forgotify" is then a handmaid to the creator's idea of a new rough category: *unlistened music*. It is these interactions between human and machine players that are now shaping our music discovery and influencing the way we categorize in Celestial Jukeboxes (that is, abundant, accessible, and connected streaming landscapes). With growth in fields of AI, we may move toward working-with-machines, rather than relying on machines to carry out simple assignable tasks (Lafrance, 2016). It is possible to imagine that machines may eventually also make rough and human-like categories of music.

Spotify lacks two very folksonomy-friendly features: tag clouds and search filters. The limitations of Spotify's text-box search render it unwieldy as a tool for browsing. The user should know what they are searching for and must be vigilant with spelling. A user will have difficulty finding a song if they are unsure of the title, as other metadata only includes artists, albums, and playlists, rather than themes, genre, lyrics, instrumentation, or other identifying factors. Instead, users must use external tools like a Google search or Shazam to gain further information. Other apps which use Spotify's API like "Forgotify" or "Every Noise at Once", provide additional discovery tools not based upon metadata such as artist and song title.

Spotify is just one service to grapple with the complexities of organizing and presenting a large array of music in a digital interface. It is palpable that Spotify promotes profes-sionally-curated content over user-made content creating both navigational ease and vocabulary limitations. User-made metadata is rarely displayed on the site, except in user profiles and user-made playlists. A controlled vocabulary is maintained by a strong presence of taxonomical features, used in wayfinding. Spotify's interface relies strongly on feature content and menus as has been shown in Figure 7.

This paper demonstrates that folksonomy can be interpreted from interfaces, but it also goes deeper than the front-end of the interface. The display of tags is an important indicator of a fully functional folksonomy, where users can add, edit, and interact with user-made music-metadata. Sites that feature visible tags support broad folksonomies, as users can adopt their referred pre-coined tags, as well as narrow folksonomies, where individuals or groups can create and use their own tags. Most importantly, the absence of tags suggests only partial folksonomy, where users may be able to add metadata but have limited functionality to edit and interact with it. Tags were not a large part of Spotify's interface. This finding evidences, at least to a degree, that Spotify's relies heavily upon automated recommendations based upon previous listening and scrollable feature content curated by a featured guest or user for music discovery. Folksonomy-friendly sites hedge their bets so to speak, by providing multiple ways to discover music and explore the interface. The lack of folksonomy in Spotify may make the interface ultimately efficient to wayfind within but at the cost of the organic movement of music folksonomies which can characterize a zeitgeist or journey a less-taken path.

Notes

1. Release Radar is a weekly personalized playlist akin to Discover Weekly, but includes only new releases and is renewed each Friday for active users.
2. Collaborative filtering compares the data of similar users to present recommendations. It is most popularly seen as "Customers who purchased this also purchased … " as found on Amazon.
3. Forgotify's Twitter bio reads "4 million songs on Spotify have never been played. Not even once. Let's change that … " See https://twitter.com/forgotify .
4. API stands for Application Programming Interface. Mulesoft Videos (2015) gives a clear and simple definition: "An API is the messenger that runs and delivers your request to the provider you're requesting it from, and then delivers the response back to you". See https://www.youtube.com/watch?v=s7wmiS2mSXY .
5. Lamere is responsible for building music recommenders for Spotify. Updates on his current work can be found on his Twitter: https://www.twitter.com/plamere; see also http://www.static.echonest.com/popcorn/.
6. https://en.wikipedia.org/wiki/Vaporwave .
7. http://everynoise.com/everynoise1d.cgi?scope = all .
8. Macdonald includes a list at the bottom of his site of interesting links that explore tags of Spotify and offer up their own discovery experience or rough category.
9. Figures 5 and 7 have been taken in Spotify in version 1.1.3.259.g8172f63a.
10. Solnit's playlist can be viewed at http://www.powells.com/post/playlist/rebecca-solnits-playlist-for-hope-in-the-dark .
11. Rasmus Fleischer associated with the Pirate Bay gave an interview on his new book on Spotify at: https://digital.di.se/artikel/han-skriver-en-bok-om-spotify-det-var-fran-borjan-en-pirattjanst.

Disclosure statement

No potential conflict of interest was reported by the author.

ORCID

Amelia Besseny ⓘ http://orcid.org/0000-0002-1059-4784

References

Anderson, C. (2010). *The long tail: How endless choice is creating unlimited demand* (p. 11). New York, NY: Random House.

Cairo, A. (2012). *The functional art: An introduction to information graphics and visualization* (p. 22). Berkeley, CA: New Riders.

Eriksson, M., Fleischer, R., Johansson, A., Snickars, P., & Vonderau, P. (2019). *Spotify teardown: Inside the black box of streaming music.* Cambridge, MA: MIT Press.

Gaffney, M., & Rafferty, P. (2009). Long tail visible: Social networking sites and independent music discovery. *Program: Electronic Library & Information Systems, 43*(4), 375–391, 377. doi:10.1108/00330330910998039

Gensler, A. (2014) *Twitter #Music pulled from app store, shutting down April 18.* Retrieved from Billboard Biz http://www.billboard.com/biz/articles/news/digital-and-mobile/5944820/twitter-music-pulled-from-app-store-shutting-down-april

Governor, J., Nickull, D., & Hinchcliffe, D. (2009). *Web 2.0 architectures: What entrepreneurs and information architects need to know* (p. 62). Sebastopol, CA: O'Reilly Media.

Hagen, A. N. (2015). The playlist experience: Personal playlists in music streaming services. *Popular Music and Society, 38*(5), 625–645. doi:10.1080/03007766.2015.1021174

Harper, R., Rodden, T., Rogers, Y., & Sellen, A. (2008). *Being human: Human-computer interaction in the year 2020.* Cambridge, UK: Microsoft Research Ltd.

Hogan, B. (2015). From invisible algorithms to interactive affordances: Data after the ideology of machine learning. In E. Bertino & S. Matei (Red.), *Roles, trust, and reputation in social media knowledge markets* (pp. 103–117). New York, NY: Springer International Publishing.

Hu, C. (2016, December 15) *What is "Escape Room" and why is it one of my top genres on spotify?* Retrieved from Festival Peak https://festivalpeak.com/what-is-escape-room-and-why-is-it-one-of-my-top-genres-on-spotify-a886372f003f

Ingham, T. (2016, August 31) *Spotify is making its own records ... and putting them on playlists.* Retrieved from Music Business Worldwide https://www.musicbusinessworldwide.com/spotify-is-creating-its-own-recordings-and-putting-them-on-playlists/

Ingold, T. (2016). *Lines: A brief history.* London, UK: Routledge.

Jeorett, P., & Watkinson, N. (2015) *From taxonomy to folksonomy: How librarians organise things.* Retrieved from YouTube https://www.youtube.com/watch?v=hCLf5xz-Dk8

Khan, I. H., & Kolay, S. (2017). Study of wayfinding behaviours in an outdoor environment. In A. Chakrabarti & D. Chakrabarti (Eds.), *Research into design for communities* (p. 174). Singapore, Singapore: Springer.

Kirn, P. (2018, September 20) *Upload music directly to Spotify: Streaming giant goes in new direction.* Retrieved from Create Digital Music http://cdm.link/2018/09/upload-music-directly-to-spotify-streaming-giant-goes-in-new-direction/

Kymäläinen, T. (2015). The design methodology for studying smart but complex Do-It-Yourself experiences. *Journal of Ambient Intelligence & Smart Environments, 7*(6), 849–860. doi:10.3233/AIS-150351

Lafrance, A. (2016) *Why do so many digital assistants have feminine names?* Retrieved from *The Atlantic* https://www.theatlantic.com/technology/archive/2016/03/why-do-so-many-digital-assistants-have-feminine-names/475884/

Lynch, P. J., & Horton, S. (1994). Interface design. In P. J Lynch, *Web style guide* (3rd Ed., pp. 95-120). London, UK: Yale University Press.

Maeda, J. (2006). *The laws of simplicity* (p. 45). Cambridge, MA: MIT Press.

Manovich, L. (2001). *The language of new media* (p. 96). Cambridge, MA: MIT Press.

Morris, J. W., & Powers, D. (2015). Control, curation and musical experience in streaming music services. *Creative Industries Journal, 8*(2), 106–122. doi:10.1080/17510694.2015.1090222

Morville, P., & Rosenfeld, L. (2006). *Information architecture for the World Wide Web: Designing large-scale websites* (3rd ed.). Sebastopol, CA: O'Reilly Media.

MuleSoft Videos. (2015). What is an API?. Retrieved from https://www.youtube.com/watch?v=s7wmiS2mSXY

Pasick, A. (2016) *Spotify is using 50,000 anonymous hipsters to find your next favorite song.* Retrieved from https://qz.com/628812/spotify-is-using-an-anonymous-army-of-50000-hipsters-to-find-hot-new-songs/

Plaugic, L. (2017) *Spotify has more than 60 million subscribers now.* Retrieved from https://www.theverge.com/2017/7/31/16070982/spotify-60-million-subscribers-july-2017

Six, J. M. (2014) *Fundamental principles of great UX design | How to deliver great UX design.* Retrieved from https://www.uxmatters.com/mt/archives/2014/11/fundamental-principles-of-great-ux-design-how-to-deliver-great-ux-design.php.

The Drum. (2016). Spotify: '*Playlists offer unique insights into a user's habits, hobbies & interests, plus unrivaled contextual targeting.* Retrieved from http://www.thedrum.com/news/2016/06/10/spotify-playlists-offer-unique-insights-user-s-habits-hobbies-interests-plus

Van Leeuwarden, B. (2018) *Music marketing (Part 1) — Artist, algorithm and audience.* Retrieved from https://www.minttwist.com/2016/11/01/the-future-of-music-marketing/

Vander Wal, T. (2005) *Explaining and showing broad and narrow folksonomies.* Retrieved from http://www.vanderwal.net/random/entrysel.php?blog=1635

Wallenberg, B. (2017) *Han skriver en bok om Spotify: Det var från början en pirattjänst.* Retrieved from https://digital.di.se/artikel/han-skriver-en-bok-om-spotify-det-var-fran-borjan-en-pirattjanst

Promises and pitfalls: the two-faced nature of streaming and social media platforms for beirut-based independent musicians

Chris Nickell

ABSTRACT

I argue that streaming services like SoundCloud and Spotify are key elements of a two-faced virtual ecosystem – including social media and crowd-funding platforms – that independent musicians world-wide must navigate to achieve economic viability of their music. The two-facedness of these technologies lies in their promotion as demo-cratic and open gateways for musicians at the same time as they actually serve as gatekeepers that throw up hidden walls along lines including national origin and language. I begin by introducing the music scenes of Beirut, Lebanon, where I have done ethnographic research since 2015. I then lay out the importance of online modes of sociality for these musicians and their fans, focusing on social media and streaming services. SoundCloud, in particular, illustrates the two-faced nature of these technologies. Throughout, I offer case studies of how several musicians and other culture workers are navigating the promise and constraints of these technologies.

Introduction

In the chilly, top-floor apartment of one of Beirut's pioneer figures of the early 2000s independent music scenes, Zeid Hamdan attaches a fresh gas tank to the propane heater in his living room with a wrench as he explains to me in English: "In this scene here, I have little popular recognition. I'm condemned to be in the underground. But abroad, there they appreciate. In Egypt, in the rest of the world. It's just at home." "Why do you think that is?," I ask. "I don't know … because I'm not cuuuute, yeah my music is not politically correct. I don't know. I'm not sexy cool. If you look at all those new bands … " He trails off. "Who Killed Bruce Lee," we say at the same time. He continues mentioning other better-known Lebanese independent bands: "Mashrou' Leila, the Wanton Bishops. Yeah they're all COOL! And I'm not cool. I'm the opposite of cool, you know? That keeps me from having the good bookings and the big crowds." "At least in Lebanon?," I suggest. "Yeah in Lebanon. Outside, I get the nice gigs because people don't see my image. They look at my SoundCloud. That's where they judge from, the sound" (Interview, January 2016). The self-theorization and slippages of ocular and auditory verbs Hamdan offers as we discuss his visual and sonic reception highlight the importance of the music streaming platform he mentions at the end, SoundCloud. Remarks like these referencing the virtual ecosystem of platforms on which musicians maintain accounts are common-place in my research, prompting inquiry into the relationship my associates have to these relatively recent but ubiquitous technological developments.

In this article, I will argue that streaming services like SoundCloud and Spotify are key elements of a challenging and often two-faced virtual ecosystem – including social media and crowd-funding platforms as well – that independent musicians across the world must navigate to achieve economic viability of their music. The two-facedness of these tech-nologies lies in their promotion as democratic and open gateways for musicians at the same time as they actually serve as gatekeepers that throw up hidden walls along lines including national origin, language, and access to funds and fast Internet speeds. I begin by introducing the music scenes of Beirut, Lebanon where I have done ethnographic research since 2015. I briefly lay out the importance of online modes of sociality for these musicians and their fans, with a special focus on SoundCloud and Spotify. SoundCloud, in particular, illustrates this two-faced nature of these technologies I seek to highlight. Throughout, I offer case studies of how several Beirut-based musicians and other culture workers are navigating the promise and constraints of these technologies.[1] I conclude by suggesting that the specific challenges they face are intensifications of the same difficulties these streaming and social media platforms are presenting for independent musicians around the world because of the fundamentally extractive logics such platforms employ.

An overview of Beirut's independent music scenes

My research associates are predominately middle-class and sub-elite male musicians active today who came of age amidst the reconstruction of Beirut during post-Civil War Lebanon of the 1990s and early 2000s.[2] They draw from diverse influences including English-language blues and folk, Arabic-language rock and vernacular poetry, and instru-mental post-rock. This article is based on ethnographic research within these music scenes, both "in real life" ("IRL") over thirteen months spanning six trips to Beirut since summer 2015 and virtually.[3] In the larger project from which this article derives, I find that independent musicians press masculinities into service to shore up social status and defer threats of downward mobility that characterize austerity regimes in the Mediterranean and beyond. Here, I focus on the specific role that the virtual ecosystem of social media and streaming services play in these "IRL" negotiations of musicians' place in Lebanese society and the global music market.

These musicians' scenes in Beirut can be considered "independent" for two main reasons. First, and ethnographically most relevant, participants in these scenes understand themselves as part of a broad and widespread set of musical practices that exist outside the imagined mainstream and can be thought of as independent, alternative, or do-it-yourself (DIY). While many musicians interchanged these and other terms in our conversations, a majority of my interlocutors referred to themselves as independent musicians. These research associates ground this self-understanding in both the rich history of popular music in Lebanon as well as their own musical omnivory fed by frequent travel abroad and the spread of digital file-sharing technologies during their formative years.[4] Second, following a popular music studies approach, the financial resources for this music are not from major labels, the large Arab media houses based in Beirut, the art music institutions of the city like the Conservatoire, or networks of folk music and poetry rooted in the working class.[5] Instead, they resemble Hesmondhalgh and Meier's observations about funding of independent music in Europe and North America today: micro-labels, venues, festivals, and professional and semi-professional musicians often work closely with local and multi-national corporations and governmental cultural foundations that generally allow more artistic freedom and intellectual property rights than the major labels, in exchange for various branding and promotional arrangements (Hesmondhalgh & Meier, 2015).[6] Many independent musicians in Beirut self-fund their musical activities, drawing from savings and family support as well as their own earnings. Funding from non-governmental organizations in Beirut through direct support of artists, venues, and festivals presents another, less-theorized financial element supporting independent music.

Scholars have characterized contemporary Lebanon as both late capitalist and thoroughly sectarianized, noting the entanglement of these two social orders.[7] Most of my interlocutors, who socially span from the middle class to just below the sectarianized elites, understand such a system to hold little promise for them. The intense reliance on social capital in order to start a business venture or even acquire and maintain a job binds them to the sectarian power structure if they hope to be able to move up the social ladder. This named reliance on stored-up and status-based social currency, called *wasta* in colloquial Levantine and Egyptian Arabic, especially disadvantages those who are professionalized through degrees in fields where few opportunities in Lebanon exist including architecture, design, literature, or art history. Many interlocutors bemoan the part-time or full-time jobs in journalism, advertising, or retail and hospitality as markers of downward mobility that nevertheless enable their continued participation in the independent music scenes they feel are an escape from the *fawḍa* ("chaos") of Beirut and Lebanon in general. Another consequence of their use of independent music to set themselves apart from the sectarianized mainstream of Lebanese society is that the potential audience within Lebanon for most bands remains restricted to others who also seek to exit that system. One manager broke it down for me, estimating that in a country of four million, only 100,000 on any given night may be inclined to go out for a drink and take in some music, so the potential crowd at any given venue with live music is small and easily exhaustible, limiting the potential for growth.

Aside from sharing this sense of escape, musicians in these scenes vary in their desires from music. Most of the artists with whom I spoke wanted to have listeners, both in live performance and through recordings, and to make a living from their music. Only a few musicians reject profit motives altogether and are content to recoup costs they incur making music. For those who wanted to succeed economically, the idea of "making a living" varied widely: some sought to cover their costs and make enough to pay rent or food, while others wanted to achieve the high standard of wealth they imagined popular musicians worldwide to enjoy. Some also voiced a desire to leave Lebanon through their musical careers, either temporarily or permanently. Facing all these musicians are constraints on resources and infrastructure that they, along with venue managers, festival organizers, grant-making organization staff, production personnel, and even listeners readily discuss with me in interviews. As the manager I referenced above pointed out, the meager potential audience for any sort of gig or more formal concert on a weeknight in the relatively small, conservative, and sectarianized country of Lebanon limits musicians' income. The lack of different sized venues with sufficient sound equipment limits the amount of shows they can put on a year. The absence of a dedicated music press and set of music management professionals requires musicians to do more promotional and logistical legwork than they feel their counterparts in global north scenes have to do.[8] These domestic problems lead many musicians to desire touring abroad, only to face the expense and visa requirements of travel that prevent many from touring as often as they would like.

Social media as a requirement for independent musicians

These desires and constraints in turn shape the ways musicians interact with social media and streaming services. Among online modes of sociality, social media have become a crucial tool for independent musicians around the world, regardless of their physical mobility, because of their reach. These platforms – including Facebook, Twitter, YouTube, Instagram, SnapChat, and SoundCloud – arose as part of a wave of new design developments through the Internet. Sometimes referred to as Web 2.0 (O'Reilly, 2005), these developments emphasize dynamic, interactive, and often user-generated content as well as accessibility. Scholars of social media have emphasized two points germane to the discussion of Beirut-based independent musicians' uses of these technologies. First, social media, like all technologies, provide *affordances* that shape but do not overdetermine the modes or content of users' interaction with them (Ellison & boyd, 2013).[9] Considering the fast-paced world of social media, Ellison and boyd make an important point:

> In order to produce scholarship that will be enduring, the onus is on social media researchers to describe the technological artifact that they are analyzing with as much care as survey researchers take in describing the population sampled, and with as much detail as ethnographers use when describing their field site (ibid. 166).

Concretely, for instance, the affordances of early 2010s Facebook before the emphasis of video content influenced my interlocutors to engage with the platform differently than mid-2010s Facebook after video content became prioritized, as described below. In a more abstract sense, Ellison and boyd's point leads me to situate and temper understandings of the promises of social media as a tool that has democratized the Internet (Kim, 2012). Social media have indeed radically opened some kinds of access to certain people who use the Internet, but this access keeps pace with contours of control built into social media platforms including algorithms to personalize content; restrictions on the configurations of privacy settings; pay-to-boost options for advertising; and featured, professionally-generated content often passed off as user-generated. These limitations operate along individual lines as well as lines of socially-maintained difference like national origin, language, and class, as I will discuss below.

Research specifically on musicians' uses of social media has tended to focus on musicians from Europe, Anglophone North America, and the antipodes. Methods have clustered around in-depth interviews (Baym, 2012; Scott, 2012; Young & Collins, 2010) and discourse analysis of social media postings and trade publications (Hracs, 2012; Meier, 2015; Morris, 2014). Scholars have found that independent musicians use social media to interact with listeners and fans to varying degrees and to advertise performances and recordings. Most articles conclude that social media offer a democratizing potential to musicians through open access platforms, but that these benefits come with drawbacks including less time for creating music and an imperative to possess or develop entrepreneurial skills on social media.

For my interlocutors in Beirut's independent music scenes, social media serve quotidian uses that line up with this existing research on independent music scenes in the global north. For instance, I often initiated contact and stayed in touch with artists through the Facebook Messenger app of the mid 2010s. They in turn use the platform to reach out to potential touring venues outside Lebanon that have Facebook pages, and sometimes they are recruited for festivals and other gigs through Messenger or, increasingly, Instagram. They also use these social media platforms simply to be social, with camaraderie and conflict playing out virtually as much as "in real life."

Alongside these rather commonplace uses of social media in Beirut's independent music scenes, it seems something more complicated is happening. Social media are no longer thought of as an optional component of independent musicians' activities. For example, several interlocutors have criticized the slapdash social media presence of Tanjaret Daghet, a Syrian rock band otherwise universally respected for their musical talents. Indeed, the band rarely updates Facebook, Twitter, SoundCloud, or YouTube with any recent activity, let alone the kinds of content tailored to each medium that savvy fans have come to expect. When I asked "So you don't like to use FB, Twitter, the like?," band member Tarek Khuluki said:

> No. On Twitter you can talk to people, FB you can organize stories … that moment that we had, which you saw, I want to *live* it more than record it. I know it's my movie. But still, doing like … do I want to remember it from here [points to his phone] or here [points to his head]? My head matters more. How did they take shape in my head? How did they cook in my head?
> (Interview, August 2017)

This rejection of social media as a poor surrogate for experiencing moments live is a sentiment shared among other musicians I work with. And yet, in light of the band's hopes to launch a crowd-funding campaign to raise money for touring abroad, critics of their absence from social media have merit: Tanjaret Daghet's lackluster online presence can be understood as an entrepreneurial failing that inhibits the band's ability to scale up their listener base.

In contrast to Tanjaret Daghet, some bands are better known for their online presentation or image than for their music. Blu Fiefer, still in the early stages of a career, has put out a strong social media presence with frequent posts, over 30,000 "Likes" on a Facebook artist's page, around 3,000 followers on Instagram and Twitter, and a professionally-designed website with several music video singles. All this activity predates the forthcoming EP. In conversation with Blu's producer, she told me "[Blu] is constantly strategizing to make her dreams a reality […] Her Instagram is a world she created for Instagram. It's not her world. Even in an interview with you she'll be in that mode" (Interview, February 2017). Blu seems to understand the importance of social media to the degree that the apparatus must be in place before leveraging the buzz around an album drop, and that it need not be reflective of reality, but that it be a reality, a world. Social media here are not auxiliary; they are essential.

Musicians in Beirut with international touring experience have expressed the same kinds of thoughts about the correlation between talent and social media presence. A guitarist who played on tour with another Beirut-based musician said in an interview:

When I went on tour […] for the first time, I was like whoa. Wow. So many good bands that are not known. But you know you watch the show and you're like oh my god who the fuck is this? And then you go to the YouTube Channel and there's like 10,000 views. [One of my own bands] has 15,000 views and we're nothing, you know what I mean? And they're playing shows in France, like amazing shows, so good. So they have the sound, the show, the clothes, they have everything.
(Interview, March 2017)

This musician is dedicated to posting regularly across social media platforms to promote his musical work. His surprise at the other musicians' low view counts reflects his understanding of social media as a necessity for musicians who desire a career. He gestures toward the discrepancy between the talent of these European bands he saw and their paltry online presence in the larger sweep of explaining how insignificant he feels he and his fellow musicians are compared with the rest of the world. And yet, at the end of the day, his band does have those view counts. This gap between the robust social media presence my interlocutor expects of a band he considers to be talented "in real life" and the reality he finds when he looks them up online suggests social media and talent come as a package. This gap between my interlocutor's expectations and reality also speaks to the increased pressure to perform well on social media facing independent musicians from smaller countries with less international mobility and weaker to non-existent state arts funding like Lebanon. Without the opportunity first to grow a large fan base "in real life," these musicians must build up social media before – or at least coincident with – their development of popularity as a live act in order to open doors for continuing to grow an audience abroad.

Producers, festival organizers, and other independent music personnel must also understand affordances of social media platforms in order to create financially viable enterprises. The organizer of the largest independent music festival in Lebanon, Wickerpark, spoke with me about the early days of the festival that coincided with Facebook's pre-boost era, where "events" were not as common and there was not a requirement to pay in order to increase visibility:

Facebook was major, major player in the success, at the beginning of the festival. Naturally. You created an event back then … and now Facebook is full of events. But before, the first two years, it's this event. That's it.
(Interview with George "Junior" Daou, February 2017; emphasis in original)

As these technologies have changed and complexified, culture workers have had to adapt. A producer of a web series and concert series called Beirut Jam Sessions, Anthony Semaan, told me about the way he thinks about Facebook and YouTube since the former began emphasizing video content in its algorithm:

So you look at someone like [Lebanese expatriate singer] Yasmine Hamdan who released her video two days ago. She put it entirely on Facebook. And she put it entirely on YouTube. Because today, each social medium is independent. You cannot promote YouTube shit on Facebook. You cannot promote Facebook shit on YouTube. Each one is independent. You need to reach as many people as possible, you need to be available on as many places as possible. You need to reach large numbers of people quickly, put your money into Facebook. YouTube is still there, it's slower. Always keep it there. Even Beirut Jam Sessions, I've been reposting old ones entirely on Facebook. Just to see what I can reach. I'm getting more traction.
(Interview, February 2017)

Both of these culture workers show how important understanding the affordances of social media platforms is to the success of music ventures given the constraints of Beirut and Lebanon for independent music.

Part of what motivates musicians to care about social media, besides the direct opportunity it provides them to engage with new potential listeners virtually, is their sense that their activity on these platforms influences their chances of being noticed by record labels. This awareness accords with my argument here that these platforms' affordances have made them into gatekeepers of the music industry, following the demise of Artists and Repertoire (A&R) divisions in the restructurings of the early 2000s. Musicians must market themselves to an imagined, virtual audience who will evaluate them and produce publicly-available metrics – Facebook likes, retweets, Instagram hearts, etc. – that music industry personnel as well as independent musicians themselves can use to negotiate resources like studio time, distribution rights, and tour opportunities.

Indeed, a conversation in September 2017 with a research associate who had just begun a job at Sony International's Middle East headquarters bore out this importance. We were speaking about the questions regarding social media that I address in this article, and he volunteered that Sony has an app that calculates reach, following, and potential of various artists that employees check every few weeks or so to see if the artists they are hoping will "make it" have crossed the threshold. He and his supervisor bring Beirut-based independent bands to their general manager's attention, only to get the response that "yeah, these guys are really great, but how are we going to make any money off of them?" This glimpse into the industry side in Lebanon confirms the ways social media have come to mediate between artist and industry as a trusted predictor of profitability through metrics like reach calculated from publicly available data on the artists' accounts across platforms.

Streaming platforms in Lebanon

Streaming services like Pandora, Spotify, Tidal, and SoundCloud have begun to receive more focused attention from scholars since 2015, owing to their outsized influence on the modes of music consumption worldwide in the past ten years. Quantitative social scientists have debated the net revenue effects of streaming services on the music industry (e.g.: Aguiar & Martens, 2016; Borja, Dieringer, & Daw, 2015; Lee, Choi, Cho, & Lee, 2016; Wlömert & Papies, 2016). Media studies scholars are investigating how Pandora, growing out of the Music Genome Project, and other Internet radio services like Songza structure "downstream" listener experiences through "upstream" algorithms and curation (e.g. Baade, 2018; Glantz, 2016; Prey, 2018). Spotify, discussed more extensively below, received sustained attention in a 2017 thematic section of *Culture Unbound* (Fleischer & Snickars, eds., 2017) and more recently in a 2018 ethnomusicological dissertation (Durham, 2018). Fleischer, Snickars, and their collaborators employ digital humanities methods to follow files and thus reverse engineer some aspects of Spotify's black-box digital music distribution platform. Durham employs virtual ethnography to understand how Spotify and a dark-web platform he calls Jekyll produce modes of sociality. This scholarship tries to get a handle on the upstream architecture of Spotify, focusing on its operations from within the virtual world. In my ethnographic research, I aim to follow the stream down into its "IRL" manifestation in the daily lives of my interlocutors. This focus downstream is one reason I will not be dwelling on the particular algorithms that structure listeners' experiences of these platforms. An additional reason is because, despite the strengths of this initial foray of research into streaming services, very little is understood about the specific design of these algorithms, likely owing to both the technical barriers to entry from the outside and the strict non-disclosure agreements company employees must sign.

Indeed, in my fieldwork, streaming platforms played an important role in how my research associates access music. Chronically slow and expensive Internet connections in Lebanon throughout the 1990s and early 2000s hampered postwar economic recovery, but it did not deter the spread of file-sharing technologies among audiophilic youth. These websites like Napster and Pirate Bay hastened the demise of La CD-Thèque, a brick-and-mortar institution of the postwar independent music scenes since the mid 1990s that often carried the only physical copies of European and North American indie music in the country. The advent of 4G wireless service in Beirut in the mid 2010s and more recently in cities in the coastal north and the south has made streaming music practical for my interlocutors, most of whom can afford data packages for their smart phones. Yet Lebanese IP addresses give far fewer choices for streaming platforms than in North America or Europe. My anecdotal evidence suggests Spotify predominates, with Lebanon-based Anghami a distant second. Friends instinctively plug in their phones when we get in the car to drive somewhere and select a few tracks they have saved on Spotify to set the mood. One prominent event producer uses YouTube mixes he makes ahead of time that include Lebanese and foreign musicians to accompany his forty five-minute drive down the mountain to the city. These uses of streaming services in everyday life of Lebanese culture workers is of a piece with such services' global saturation over the past five to ten years.

While listeners in Lebanon face limited choices of streaming platforms, musicians aim to upload their music to as many aggregators as possible in order to ensure it appears in the global databases of services including Spotify, Pandora, Google Play, and Anghami that serve Beirut-based musicians' potential audiences abroad.[10] Yet artists have little control over the popularity of their tracks once they are uploaded to these services because algorithms determine track lists for different listeners based on individual listening history, others' listening histories, and musical characteristics of a given track. Evidence from speaking with musicians in my fieldwork suggests that breakthroughs on these platforms are rare, and the causal relationship between success on streaming services and a band's reach internationally remains unclear. One anecdote puts the random element of algorithmic luck into perspective. In speaking about sources of revenue, Beirut-based English-language dream-pop trio Postcards recounted how "Walls," a song from their 2015 folk-rock sophomore EP, made it onto a popular Spotify playlist. Lead vocalist and lyricist Julia Sabra explained:

> From online stuff, we've made like $3,000 so far. Which is unexpected because it's all streaming and streaming makes like 0.0 cents. It's all streaming. We've probably made like $200 selling music online and the rest is streaming. I don't know how because, like, "Walls" got on one of Spotify's playlists, like folk something, and it got like 50,000 listens. But it doesn't happen a lot. No, it's great, we're making money off of it.
> (Interview, March 2017)

A little later in our conversation, discussing European countries in which they have a touring presence, Sabra and guitarist Marwan Tohme explain another positive consequence of this exposure:

JS: Italy came kind of randomly. Paolo, [our contact there,] heard "Walls" on like Spotify and was like "Can you guys come to Italy?" This is how it all, so it wasn't … Germany, he heard Walls too –

MT: – and he asked Paolo about us. Fadi [our producer] was just telling me, he was skyping with him and he told him.

(ibid.)

Through steady development of original repertoire and online presence, along with a crucial lucky break in Spotify's playlist algorithm, Postcards have built a modest touring presence in Portugal, Italy, and Germany that covers their costs and earns a bit of money. Because artists have so little knowledge of – let alone control over – the ways these streaming services prioritize tracks for certain listeners, most of the effort of Beirut-based independent musicians is spent on platforms that have social elements they can more directly curate. It is in the hybrid social-streaming platform of SoundCloud that we can observe how systemic inequities for independent musicians worldwide that inhere in the virtual ecosystem become magnified along lines of race, language, and social class that further disadvantage independent musicians working outside the global north.

Soundcloud: a social-streaming hybrid

SoundCloud is an audio-sharing social media platform launched in 2008 that has grown to host over 100 million tracks with over 175 million monthly users as of the latest statement in February 2016. Since I began my research in 2015 and as of this writing in 2018, it has provided "free" and monthly fee-based "Pro" and "Pro Unlimited" account options tailored to listeners and small-scale musicians, and to semi-professional and professional musicians, respectively. The general "About" statement on SoundCloud's website offers a glimpse of these systemic inequities and is worth quoting in full:

> As the *world's* largest music and audio platform, SoundCloud lets people discover and enjoy the greatest selection of music from the most diverse *creator community* on *earth*. Since launching in 2008, the platform has become renowned for its unique content and features, including *the ability to share music and connect directly with artists*, as well as *unearth breakthrough tracks*, raw demos, podcasts and more. This is made possible by an open platform that directly connects creators and their fans *across the globe*. Music and audio creators use SoundCloud to *both share and monetise their content with a global audience*, as well as receive detailed stats and feedback from the SoundCloud community. (About SoundCloud, n.d., accessed November 2018; emphasis mine)

Taking this text as a mission statement of sorts, I suggest SoundCloud seeks to attract people to the site through the community ethos ("creator community") and emergent aesthetics ("breakthrough tracks") that have long characterized independent music, even more so since the advent of social media. The use of words like "world," "earth," and "globe" alongside "diverse" convey a totality and heterogeneity that SoundCloud seeks to telegraph. This totality and heterogeneity hearken to two items in an academic lexicon: "globalization," a conception of the world as an integrated system; and "neoliberalism," a regime for, among other things, managing difference. SoundCloud's participation in the extractive nature of that regime becomes clear to the reader in the final sentence through the reference to "monetis[ing] content [sic]." Indeed, culture industries have been specifically targeted for monetization in the past two decades with the rise of web-based technologies.

SoundCloud in particular extracts revenue through several methods that are worth detailing. From artists with more than three hours of audio uploads, the website collects a monthly fee for Pro and Pro Unlimited accounts. From corporations seeking advertisements, SoundCloud collects fees for arranging many levels of access including branded competitions, playlists, and artist exclusives. From listeners, SoundCloud recently began collecting monthly fees for ad-free and off-line access to "the world's largest streaming catalog" through SoundCloud Go and Go+, directly competing with other streaming services including Spotify and Pandora. SoundCloud pays artists they invite to join their Premier level a share of the revenue their tracks generate monthly. They have also brokered deals with major content owners including labels and publishers to supply "premium" content for their new streaming services. These moves effectively position SoundCloud to capitalize on the ethical and aesthetic formation we could call the indie vibe, collecting revenue from various sources but paying only the most successful of their "holdings." This arrangement replicates the exclusivity of the "old" music industry, disadvantaging the vast majority of emergent music-makers they claim to champion. SoundCloud thus serves as both gateway and gatekeeper, emphasizing the former role's promise while extracting resources under the rubric of the latter.

How, then, do Beirut-based musicians engaging with SoundCloud and other social media handle these necessary but extractive environments? I close with several examples of research associates who navigate material and psychic investments in SoundCloud in ways that speak to both its promise and its limitations. First, a key affordance of SoundCloud with which my research associates have engaged is the ability for artists to see not only total "listen" counts for each track, but from which countries their listeners are accessing SoundCloud's servers. Pro users see a breakdown by cities as well. This feature came up prominently during a January 2016 interview with Fer'et Aa Nota, a group of six twenty-something musicians, all with varying levels of conservatory training, who incorporate elements of jazz and ṭarab (Arab art music) in a self-styled Oriental Fusion. They are currently working on their second studio album and gig occasionally throughout Lebanon at festivals and one-off shows, never having performed outside the country. All of the musicians have day jobs to cover basic expenses, with their free time dedicated to managing the band and making music together.

While we were discussing their career plans, the saxophonist explained that they were in contact with a booking agency that works with other regional acts based in Egypt in the hopes that they can arrange a show there sometime soon. He said during our interview: "According to our social media insights, from SoundCloud, Egypt is our number one country [...] not Lebanon." When I replied, "Well, I guess that would make sense, 70 or 80 million people vs. 4 million here," he explained that he actually thought it was for a different reason: "Compared to other Arab countries, Egyptians are still interested in alternative music and the indie scene." He discussed how he feels there is also an appreciation of heritage [turāth] or culture [thaqāfa] in Egypt which is not present in Lebanon given their generation's prevailing interest in Arabic and Western contemporary popular music. The saxophonist's younger brother, their social media manager, chimed in to say that "In Lebanon, if you have Arabic lyrics, it's mainly bleah, you're boring. In Egypt, it's culture" (Interview, January 2016).[11] SoundCloud's listener breakdown by country thus allows this

Oriental fusion band to imagine an audience in Egypt that they began actively working to court based on that information. They have even theorized explanations for this trend based on socially-informed stereotypes of language ideologies in both countries to explain their apparent popularity in Egypt and their relative lack of support from the hometown crowds. Egypt, they imagine, could provide an audience that would embrace their artistic values of lyrical sincerity and a hybridized but "authentic" Arabness while supporting them economically through attendance at their imagined live performances.

And yet there are structural limitations to the enactment of the fantasies this affordance enables. For one, Egypt no longer supports the kind of alternative music scenes that this band imagines playing for (Sprengel, 2017). For another, the country count feature is marketed to artists under an assumption of mobility that ignores limitations on artists from outside Europe and North America, especially those from the Middle East.[12] SoundCloud quotes U.S.-based rapper GoldLink in its advertisement of this feature, saying "Country and city stats are ridiculously cool. While we planned this upcoming European tour we were able to see exactly which cities responded to the music, and we made sure to get there" (On SoundCloud - Pro, n.d.). Even with months of advanced planning and visa fees paid upfront, artists based in Beirut sometimes do not receive permission to travel in time for their planned tours. These material limitations did not inhibit this band from using SoundCloud's country count to help imagine an Egypt tour, marshaling energy toward a goal that, as of the time of writing, has yet to be attained as the band struggles to find the resources to make new music and book gigs locally.

A second case study involves a prominent group of roughly six Beirut-based, Arabic-language rappers who use SoundCloud to host their tracks, both one-offs and albums. They link to these tracks on straight-ahead social media platforms like Facebook to announce releases and promote their events. Much can be said of the ways these rappers' release their tracks, for free and often as they finish them, that speaks to their distinct values apart from others in the independent music scenes. Most relevant for my purposes here is the way they occupy space on the SoundCloud platform. By using the platform, they are reaching beyond the limited "IRL" spaces they occupy in and around Beirut, but they are not gesturing toward a "global audience" or a shared "SoundCloud community" mentioned in the mission statement above. Instead, by using almost exclusively Arabic-language text written in Arabic script, these artists have created a space larger than their physical reach but not accessible to the whole site. Indeed, most of these artists will not appear in a search of their transliterated names, and most track names contain little to know Latin script. They are only accessible if a listener clicks through from a social media platform – where Arabic text also predominates – or knows what she is looking for on the SoundCloud search engine.

Among the effects of this tactic are the creation of a closed-off community within the posited openness of the "SoundCloud community." Only those listeners with the language skills to understand these rappers' music will find it. As a result, these artists have between 5,000 and 20,000 followers, mostly Arabs from across the Middle East and North Africa and in the diaspora, a core group of whom engage regularly with the tracks to provide encouragement and feedback in response to elements that move them.[13] All these rappers except two use the free capacity of three hours' upload time.[14] Nearly all of their commenters' accounts are the basic version, too. In this way, I suggest these rappers have solved the bind of the two-faced social-streaming hybrid SoundCloud for themselves by refusing to use it for its intended purposes of growing a listener base that drive artists to pay the fees. Instead, by using Arabic language and uploading modest amounts of content, they are using the service to share music among a closed-off community defined by listeners' abilities to understand the music linguistically and, based on affirmative comments, politically. This non-participation in the profit motives of SoundCloud within the small circle of artists represents an important tactic for navigating these technologies' promise and limitations.

In concluding, I return to Zeid Hamdan, the singer-songwriter and producer from my opening anecdote, to show how he makes SoundCloud work for his needs. The site's clean site design for artist pages emphasizes the sonic over the visual, though the site coaches artists to include quality visuals as well. The focus on the visual representations of the track's sounds – as amplitude waves over time – stand in for the saturation of photographic media and/or text in social media platforms like Facebook, Twitter, and YouTube foreground sound. This sonic emphasis of SoundCloud fits Zeid Hamdan's needs. A self-professed perfectionist, he says he will not release a track "unless I've heard it a thousand times." The tight rhythms, minimal instrumentation, and careful mixing indeed give his trip-hop tracks a durability that listeners I have spoken with appreciate.

His focus on sonic craft comes at the expense of care for his image, which SoundCloud conveniently marginalizes in its emphasis on the sonic and visual representations of it. In the opening quotation, Hamdan theorizes his own current position in the Beirut scenes as marginal because of this lackluster image. In a city where image matters quite a lot, he has a point. At home, he largely performs in marginal alternative festivals. In his self-theorizing he is able to imagine through the medium of SoundCloud that his music circulates much more readily and widely than his image, reaching potential audiences in the Gulf, Egypt, and Jordan where he is asked to perform live most often these days. Again, we see the role the virtual plays in enabling not just imagined but physical mobility.

The platform of SoundCloud, then, provides certain affordances and places certain limitations on the user which constrain but do not determine their modes of interaction with it or the knowledge such interaction generates. Hamdan's conception of his audiences and, in turn, his conception of "success" or "making it" are shaped by the reliance on this medium and his own ways of engaging with it, in particular the de-emphasis on image that frees him to continue focusing on music at the expense of the visual presentation. Hamdan's definition of success, similar to those of other musicians I considered above, is two fold: to be able to live off music alone, economic stability; and to have people know him and appreciate his music widely, popularity. He has achieved the economic goal by diversifying as a film score and jingle composer, but popularity here eludes him in his own theorizations precisely because he refuses to focus on his image. SoundCloud enables him to continue ignoring the image, reinforcing a sacrifice he seems willing to make.

Conclusion

In this article, I have offered several examples of contemporary Beirut-based independent musicians interacting with the two-faced nature of social media and streaming platforms. My interlocutors are not exceptional. Instead, their situation coming from a small country with little audience for independent music and no state arts funding, and with a passport that grants less mobility than the passports of other nation-states, intensifies the promises and pitfalls of these platforms. The operators of these platforms claim to embrace openness and democracy, but they are driven first and foremost by profit motive, which that rhetoric of openness and democracy serves. This profit motive leads these platforms to extract value from as many sources as possible, including from the generators of content, the artists themselves. These platforms create hierarchy within them that tends to reinforce – rather than ameliorate – hierarchies of positionality that exist "IRL." Such hierarchies are not a bug, but a feature necessary for the systematic extraction of value from participants. Spotify both negotiates pay rates for given tracks and codes the algorithms that determine the likelihood that track will be chosen to play for a given set of listeners, giving them immense control over the fates of artists on the platform. SoundCloud can charge monthly or annually for hosting space above a certain length of upload time and can also monetize both sides in the pairing of artists with audio uploaded to the site and corporations seeking advertising material. These hierarchies generate value for the platforms, both in terms of financial and cultural capital. The return for artists is diminished for my interlocutors as it gets refracted through their circumstances in Lebanon, but very few independent musicians worldwide are getting returns that line up with the value these platforms extract from them. And yet, my interlocutors put this infrastructure to creative (mis)use for themselves in clever ways that resonate with the approaches of independent musicians worldwide. Postcards parlays the unexpected featuring of "Walls" on Spotify into small tours in Germany, Italy, and Portugal. Hip hop artists hang out in an Arabic-language corner of SoundCloud unmolested by the sea of Anglophone DJs vying for listen counts. And Zeid Hamdan continues to focus on the sound of his music rather than his lackluster image because his main portal to the outside world, his SoundCloud page, allows him to do so. Even as these platforms serve out a true function as gatekeepers more so than the gateways they claim to be, independent musicians from the global South and the north often find ways of skirting those walls by taking advantage of those same platforms' affordances to meet their needs.

Notes

1. I use the framing of "culture workers" here to include producers, promoters, band managers, DJs, venue managers, festival organizers, non-profit foundation workers, multinational corporation branding and advertising workers, and others involved in the production and dissemination of independent music in Lebanon. This framing owes much to Bourdieu's formulation of "cultural intermediaries" in *Distinction* and the subsequent academic literature considering the roles and class positions these figures occupy (see, in particular: Bourdieu, 1984; Beck, 2005; Negus, 2002;; Wright, 2005). All of these figures play a role in the virtual and physical manifestations of independent music in Lebanon, so I choose to group them under this rubric of "cultural workers."
2. Those of my interlocutors who have been active since the mid 1990s, the "old guard" among the independent musicians today, tend to narrate the scene beginning with themselves and the infrastructure of studios, labels, venues, and bands they built up. However, the history of independent music in Beirut extends at least as far back as the pre-war years. Even at the height of the war, some research associates recournted, there were politically-committed independent bands at the American University of Beirut, a couple venues for rock and for dancing, and a radio station playing this kind of music.
3. All attributed quotations have been approved in context by my research associates who offered them in interviews per my Institutional Review Board (IRB) protocol filed with the New York

University IRB as IRB-FY2016-1237. This process has also yielded valuable feedback in the writing of this article that I cite where applicable. I have sought approval for attributed quotations of non-public social media postings, anonymizing them wherever approval was not obtained. I was not able to dialogue with research associates about my use of all publicly-posted materials from which I draw, although these conversations are an integral and ongoing part of my research method.

4. Even as my research associates almost universally appreciate the music of popular and politically-engaged artists like Ziad Rahbani and Marcel Khalifeh whose careers peaked in the 1970s and 1980s, they set themselves apart from this generation of Lebanese music just as they set themselves apart from the broader Lebanese society which these older musicians engaged and critiqued. As I explain below, many of my interlocutors come from a de-politicized sub-elite and middle class background, which informs their desire to stand wholly apart from what they identify as the social ills of a sectarianized Lebanon.

5. These folk practices involve mostly wedding music and the sung poetic form of *zajal* where men improvise in a competition of verbal wit. Independent musicians I work with occasionally draw on these forms in their work, as in a 2017 ElectroZajal event, sponsored by Red Bull Lebanon, in which six DJs provided electronic backing to *zajal* tracks. See Chapter Two of [Author's Name] 2019 for more details. Otherwise, these spheres of folk production and consumption remain separate from the ones I focused on during my fieldwork.

6. See also Carah (2010) for relationships between independent music and brands in Australia and Meier (2017) for a more global consideration of the role of brands in music.

7. See, in particular: Joseph (1983), Traboulsi, 2016), Deeb & Harb, 2013), Salloukh, Barakat, Al-Habbal, Khattab, and Mikaelian (2015), and Baumann, 2017). I follow many of these scholars in using the adjective "sectarianized" rather than "sectarian" because it conveys the processural and intentional nature of the dominant social order in Lebanon: sectarianism is neither natural nor inevitable, but rather has been a conscious political strategy deployed by the ruling elites to legitimize their power.

8. These sentiments around lack of infrastructure do, in fact, resonate across independent music scenes worldwide, including in the global north. Garland (2014) describes similar feelings among musicians in both Chile and Brazil, while Luvaas (2012) discusses indie music and design in Bandung and Yogyakarta, Indonesia. Fournet (2019) focuses on these feelings among women producers in the U.S. and Peru.

9. See also Kirschenbaum (2008), Chun (2011), and Sterne (2012), among others for prominent statements of a similar argument about other forms of digital technology.

10. It bears mentioning that this uploading process is often less smooth than it sounds and requires advanced planning. Internet upload speeds are still slow, even in metropolitan Beirut, and many companies impose daytime capacity limits. My interlocutors often will plan to stay up late to take advantage of the nighttime unlimited capacity and higher upload speeds to shepherd new audiovisual content onto various platforms.

11. The topic of language ideology among independent musicians in Lebanon and Egypt is beyond the scope of this paper. It will suffice to say that a far greater propotion of independent musicians in Egypt sing in Arabic than in Lebanon. The simplest explanation for this reality, in keeping with some themes of this paper, is that Egypt has a large enough population of middle-class and sub-elite listeners to support a domestic scene in Egyptian Arabic. See Sprengel (2017) for a detailed study of this scene. More complicated explanations for the prevalence of Arabic in Egyptian independent music and its relative absence in the Lebanese scene stem from the different ways the Egyptian and Lebanese bourgeoisie have navigated European colonialism and its aftermath.

12. One of the only other scholarly publications about SoundCloud I could find – Allington, Dueck, and Jordanous (2015) – focuses on the importance of place in the valuing of electronic music on the social streaming site. From quantitative networking analysis, they find that artists not hailing from London, New York, and Los Angeles face similar levels of geographic inequality in the supposedly ubiquitous virtual space as in cultural economy taking place in physical space.

13. The other scholarly publication involving SoundCloud performs an analysis of English-language comments scraped from the platform in 2013 to investigate the modes of sociality the site enables (Hubbles, McDonald, & Lee, 2017). They conclude that most comments are meant as "broadcasting" of support or disapproval instead of invitations to dialogue. Anecdotally, I can attest that a much higher percentage of comments on these Arabic-language hip hop tracks are in the spirit of dialogue.
14. One of the two Pro users, Chyno, pays for the service to increase exposure for his English-language material, and the other, El Rass, does so because he needs the extra space for all his material.

Acknowledgments

I am grateful to Jane Sugarman and her Music & Mobilities seminar for the support to begin thinking critically about the roles of social media and streaming services in Beirut's independent music scene. I thank the American University of Beirut's Media Studies faculty for the opportunity to present this material at their January 2017 *Rethinking Media Studies through the Middle East* conference. Last, I give gratitude for Grace Osborne's true friendship, intellectual camaraderie, and sharp editing eye.

Disclosure statement

No potential conflict of interest was reported by the author.

References

About SoundCloud. (n.d.). Retrieved from https://soundcloud.com/

Aguiar, L., & Martens, B. (2016). Digital music consumption on the internet: Evidence from clickstream data. *Information Economics and Policy, 34*, 27–43. doi:10.1016/j.infoecopol.2016.01.003

Allington, D., Dueck, B., & Jordanous, A. (2015). Networks of value in electronic music: SoundCloud, London, and the importance of place. *Cultural Trends, 24*(3), 211–222. doi:10.1080/09548963.2015.1066073

Baade, C. (2018). Lean back: Songza, ubiquitous listening and Internet music radio for the masses. *Radio Journal: International Studies in Broadcast & Audio Media, 16*(1), 9–27. doi:10.1386/rjao.16.1.9_1

Baumann, H. (2017). *Citizen Hariri : Lebanon's neoliberal reconstruction*. New York, NY: Oxford University Press.

Baym, N. K. (2012). Fans or friends?: Seeing social media audiences as musicians do. *Participations, 9*(2), 286–316.

Beck, A. (Ed.). (2005). *Cultural work: Understanding the cultural industries*. New York, NY: Routledge.

Borja, K., Dieringer, S., & Daw, J. (2015). The effect of music streaming services on music piracy among college students. *Computers in Human Behavior, 45*, 69–76. doi:10.1016/j.chb.2014.11.088

Bourdieu, P. (1984). *Distinction: A social critique of the judgement of taste (R. Nice, trans.)*. Cambridge, MA: Harvard University Press.

Carah, N. (2010). *Pop Brands: Branding, popular music, and young people*. New York, NY: Peter Lang.

Chun, W. H. K. (2011). *Programmed visions: Software and memory*. Cambridge, MA: MIT Press.

Deeb, L., & Harb, M. (2013). *Leisurely Islam: Negotiating geography and morality in Shi'ite south Beirut*. Princeton, NJ: Princeton University Press.

Durham, B. (2018). *Regulating dissemination: A comparative digital ethnography of licensed and unlicensed spheres of music circulation* (Ph.D. Dissertation). University of Oxford.

Ellison, N. B., & boyd, D. M. (2013). Sociality through social network sites. In W. H. Dutton (Ed.), *The Oxford handbook of internet studies* (pp. 151–169). New York, NY: Oxford University Press.

Fleischer, R., & Snickars, P. (2017). Discovering spotify – A thematic introduction. *Culture Unbound: Journal of Current Cultural Research*, 9(2), 130–145. doi:10.3384/cu.2000.1525.1792130

Fournet, A. (2019). *Women music producers: Sonic innovation from the periphery of a male-dominated industry* (Ph.D. Dissertation). New York University, New York.

Garland, S. (2014). *Music, Affect, Labor, and Value: Late Capitalism and the (Mis)Productions of Indie Music in Chile and Brazil* (Ph.D. Dissertation). Columbia University, New York, NY.

Glantz, M. (2016). Internet radio adopts a human touch: A study of 12 streaming music services. *Journal of Radio & Audio Media*, 23(1), 36–49. doi:10.1080/19376529.2016.1155124

Hesmondhalgh, D., & Meier, L. M. (2015). Popular music, independence and the concept of the alternative in contemporary capitalism. In J. Bennett & N. Strange (Eds.), *Media independence: Working with freedom or working for free* (pp. 94–116). New York, NY: Routledge.

Hracs, B. J. (2012). A creative industry in transition: The rise of digitally driven independent music production. *Growth and Change*, 43(3), 442–461. doi:10.1111/grow.2012.43.issue-3

Hubbles, C., McDonald, D. W., & Lee, J. H. (2017). F#%@ that noise: SoundCloud as (A-)social media?. *Proceedings of the Association for Information Science & Technology*, 54(1), 179–188. doi:10.1002/pra2.2017.14505401020

Joseph, S. (1983). working-class women's networks in a sectarian state: A political paradox. *American Ethnologist*, 10(1), 1–22. doi:10.1525/ae.1983.10.1.02a00010

Kim, J. (2012). The institutionalization of YouTube: From user-generated content to professionally generated content. *Media, Culture & Society*, 34(1), 53–67. doi:10.1177/0163443711427199

Kirschenbaum, M. G. (2008). *Mechanisms: New media and the forensic imagination*. Cambridge, MA: MIT Press.

Lee, M., Choi, H., Cho, D., & Lee, H. (2016). Cannibalizing or complementing? The impact of online streaming services on music record sales. *Procedia Computer Science*, 91, 662–671. doi:10.1016/j.procs.2016.07.166

Luvaas, B. (2012). *DIY style: Fashion, music and global digital cultures*. London, U.K: Berg.

Meier, L. M. (2015). Popular music making and promotional work inside the "new" music industry. In K. Oakley & J. O'Connor (Eds.), *The Routledge companion to the cultural industries (pp. 402–412)*. New York, NY: Routledge.

Meier, L. M. (2017). *Popular music as promotion: Music and branding in the digital age*. Malden, MA: Polity Press.

Morris, J. W. (2014). Artists as entrepreneurs, fans as workers. *Popular Music and Society*, 37(3), 273–290. doi:10.1080/03007766.2013.778534

Negus, K. (2002). The work of cultural intermediaries and the enduring distance between production and consumption. *Cultural Studies*, 16(4), 501. doi:10.1080/09502380210139089

O'Reilly, T. (2005). What is web 2.0. n.p.: O'Reilly Media, Inc. Retrieved from https://www.oreilly.com/pub/a/web2/archive/what-is-web-20.html

On SoundCloud - Pro. (n.d.). Retrieved from https://on.soundcloud.com/pro

Prey, R. (2018). Nothing personal: Algorithmic individuation on music streaming platforms. *Media, Culture & Society*, 40(7), 1086–1100. doi:10.1177/0163443717745147

Salloukh, B. F., Barakat, R., Al-Habbal, J. S., Khattab, L. W., & Mikaelian, S. (2015). *Politics of Sectarianism in Postwar Lebanon*. London, U.K.: Pluto Press.

Scott, M. (2012). Cultural entrepreneurs, cultural entrepreneurship: Music producers mobilising and converting Bourdieu's alternative capitals. *Poetics*, 40(3), 237–255. doi:10.1016/j.poetic.2012.03.002

Sprengel, D. (2017). *'Postponed Endings': Youth music and affective politics in post-uprisings Egypt* (Ph.D. Dissertation). Los Angeles, CA: University of California, Los Angeles.

Sterne, J. (2012). *MP3: The meaning of a format*. Durham, NC: Duke University Press. doi:10.1215/9780822395522

Traboulsi, F. (2016). *Social classes and political power in lebanon.* Beirut, Lebanon: Heinrich Böll Stiftung - Middle East Office.

Wlömert, N., & Papies, D. (2016). On-demand streaming services and music industry revenues — Insights from Spotify's market entry. *International Journal of Research in Marketing, 33*(2), 314–327. doi:10.1016/j.ijresmar.2015.11.002

Wright, D. (2005). Mediating production and consumption: Cultural capital and 'cultural workers.'. *The British Journal of Sociology, 56*(1), 105–121. doi:10.1111/bjos.2005.56.issue-1

Young, S., & Collins, S. (2010). A view from the trenches of music 2.0. *Popular Music & Society, 33*(3), 339–355. doi:10.1080/03007760903495634

Beyond the black box in music streaming: the impact of recommendation systems upon artists

Marcus O'Dair and Andrew Fry

ABSTRACT

As algorithms have emerged as a key site of power in contemporary culture and society, they have been scrutinised by a number of media scholars, variously focusing on their opacity, bias and social implications. There has also been important work calling for a shift of attention away from the algorithms themselves to the actors that control them: the fundamental questions we should be asking of algorithms, after all, concern more than the specifics of code. This paper applies the arguments developed by Gillespie and Bucher to the algorithms utilised by music streaming services – the powerful but opaque curatorial systems that suggest songs to users. Although there has been important work on algorithms in the context of music streaming, this focus on music streaming remains relatively unusual. Even in the context of music streaming algorithms, our approach is also novel, in that we focus not on the possible effects upon *users* of music streaming platforms – that is, music fans – but, rather, on the possible effects on music *creators*. What, then, might be the effects upon songwriters and artists of the increasing prevalence of recommendation systems in music streaming?

Introduction: the growth of music streaming

Although the terms are commonly conflated, this article concerns the *record* industry as distinct from the broader *music* industry; our focus is those aspects of the music industry that are directly related to recorded music. Opinions differ as to the extent of the shift experienced by the record industry since the emergence of digital technology. On the one hand, Wikström (2013, p. 4) identifies a "dramatic" transformation, characterised by high connectivity and reduced control; music as a service, rather than a product; and an increase in amateur production. On the other hand, Rogers (2013, p. 178) insists that claims of a digital revolution are overstated: "forces of change have been diluted by forces of continuity". Wikström and Rogers disagree primarily on the effects of digital disruption on record industries and artists. One particular change in the record industry, however, is beyond doubt: streaming platforms, such as Spotify, Apple Music, Tidal, Deezer and YouTube, have become increasingly dominant, ushering in a shift from ownership to access.

Globally, income from music streaming increased by 41.1% in 2017 from the previous year (IFPI 2018). This growth in streaming income, particularly from paying subscribers rather than those non-paying users accessing ad-supported streams, is the critical factor in an 8.1% increase in income from recorded music overall; revenue from downloads in the same period fell by 20.5%, and revenue from physical sales fell by 5.4% (IFPI 2018). Digital revenues now account for 54% of the global recorded music market, and streaming has for the first time become the single largest recorded music revenue source (IFPI 2018). This increase is particularly significant because it marks the third consecutive year of growth for recorded music globally – and because those three years of growth follow 15 years of decline (IFPI 2018). Working on behalf of record labels, the International Federation of the Phonographic Industry (IFPI) accuses certain streaming platforms, notably YouTube, of a "value gap": a "mismatch between the value that some digital platforms … extract from music and the revenue returned to the music community – those who are creating and investing in music" (IFPI 2018, p. 26). Yet streaming, more broadly, has been welcomed with open arms by the record industry, as the first good news since Napster ushered in the era of peer-to-peer file sharing around the turn of the millennium. By the end of 2017, music streaming platforms had 176 million paying subscribers globally, and the pace of growth is high: 64 million of those subscribers only signed up during that year (IFPI 2018, p. 10). Although paid-for streaming subscriptions might be reaching saturation in certain markets, growth is expected in other countries including Germany, Japan and Brazil (Mulligan, 2018a). Spotify and Apple Music are the big two premium services worldwide, as of November 2018, with 87 million (Aswad, 2018) and 50 million (Chang, 2018) premium subscribers respectively. YouTube and SoundCloud actually have more users but, as primarily free services, they generate lower revenues. We can also identify some geographical variation. Apple Music tends to be more prominent in the United States, for instance, while Spotify dominates in in Europe. Other services are prominent in other territories: Deezer in France; QQ Music and Netease Music in China; Yandex and VK in Russia; Saavn in India. It is important to remember that total income from recorded music is still relatively low if viewed from a broader historical perspective: 'total industry revenues for 2017 were still just 68.4% of the market's peak in 1999' (IFPI 2018, p. 10). It does seem, however, that streaming, in 2018, is the only game in town for recorded music: despite media claims of a "vinyl revival", global revenue from physical sales was lower in 2017 than it has been at any point this millennium (IFPI 2018). In the specific United States context, for example, streaming accounted for 75% of recorded music revenues (physical 10%, downloads 12%, sync 3%) during the first half of 2018, accounting for $3.4 billion in revenue – up from $1 billion in the first half of 2015 (RIAA, 2018).

Services such as Spotify and Apple Music have tens of millions of users, and provide tens of millions of songs. To manage systems of this scale, streaming services utilise a wide variety of technologies, including algorithms. Contemporary listeners do not only rely on services' front-end interface to find music; increasingly, they also turn to suggestions from the services themselves – relying on service-generated playlists or radio stations to discover individual songs. The decline in the importance of the album is nothing new: the process of "unbundling" albums into single tracks began with the launch of iTunes. What *is* new is the fact that individual songs are now "pushed" to listeners by playlists, rather than being actively sought out. This is part of a broader trend, in which algorithms have become "a key logic governing the flows of information on which we depend" (Gillespie, 2014, p. 167). Consider, for instance, Netflix's decision to license the *House of Cards* television series, starring Kevin Spacey, in 2011. While more traditional networks considered *House of Cards* unpromising, Netflix was able to analyse the viewing habits of its 33 million subscribers – and to conclude that there was, in fact, a sufficient potential audience to commission the show (Smith & Telang, 2017). The series became a significant hit. Yet customer data did more than allow Netflix to commission *House of Cards* when others could not see the potential. Netflix was also able to promote the series in a novel fashion, for instance by creating multiple trailers for the show. Subscribers who had liked Kevin Spacey's previous work were shown one trailer; subscribers known to like programmes with strong female lead characters were shown a different trailer; subscribers known to have enjoyed the previous work of director David Fincher were shown a third (Smith & Telang, 2017, p. 8). Netflix, in other words, had identified "a new way to promote content (through personalised promotional messages based on individual preferences)" (Smith & Telang, 2017, p. 11). Music streaming services use data in a similar fashion, in the process establishing themselves as a new kind of intermediary: gatekeepers providing an interface between listener and catalogue – and acting, in the process, as DJ, radio station, storefront, and following Spotify's October 2018 investment in DistroKid (Moon & Nellis, 2018), distributor all in one.

Streaming services all have various tools to aid users in music discovery, and the balance between human and machine input varies from one to the next. These include: *text search*, for artist, release, genre, playlist name; *categorical lists*, grouping albums, artists and playlists by themes (for example Spotify includes "moods", "decades", "focus", "gaming") or outlining new releases; *playlists*, including by genre, mood, theme, popularity charts (by territory); and *hyperlinks* – any artist or album name is a clickable link to other parts of the catalogue. The user interface of these systems is critical in determining how users interact with the music catalogue. For example, an artist's "latest release" is shown at the top of their Spotify profile, followed by a list of five "popular" songs (not necessarily in descending order of streams). Priority, then, is given to the new and popular – and this is largely beyond the artist's control. The "latest release" section, incidentally, is to some extent a misnomer; if an artist hasn't released music for a certain period, "latest release" disappears and "popular" rises to the top of the artist's profile page. Design choices made by music streaming services also affect discovery using the search bar. Results for searches via text strings – for example an artist name – will vary by user. Searching "James" may give one user James Blake as the top result, for instance, and another user James Brown. There are, then, subtle systems in place that push listeners toward certain artists or songs, and away from others. As listeners increasingly rely on service's tools to find music, it becomes increasingly important for artists to be noticed, or made visible, by those tools in order to maximise their share of streaming revenue. Understanding the systemic tools is therefore critical for recording artists, and this is true for the most established through to the least known. The prominence of streaming services has made the methods by which listeners find music critical to artist's business models. Just as appearing on the first page of a Google search has long been key for online discovery, the logic emerging on music streaming platforms is as simple as it is ruthless: if an artist's catalogue is not easily discovered, she will miss out on an audience and, therefore, revenue.

As the use of algorithms by music streaming services has gained prominence, so has it come under scrutiny by scholars. On the whole, those who have examined the impact of algorithms tend to focus on the effects on users. Prey (2017), for instance, has examined the ways in which individual listeners fall into categories designated by advertisers, arguing that commercial imperatives shape "algorithmic individuation" on music platforms such as Pandora Internet Radio and Spotify. Meanwhile, Airoldi, Beraldo, and Gandini (2016) have examined the network associations within YouTube music videos, arguing that traditional genres increasingly sit alongside new algorithmically generated categories – "music for relaxation", for instance, is as effective as "rock" within machine categorisation. Our focus in this article, however, is the impact of algorithmic power – understood as located not in the algorithms themselves but in the actors behind those algorithms – on *artists*. Such a focus might seem odd: as Stahl (2013, p. 2) acknowledges, "successful artists often appear to us as paragons of autonomous self-actualisation". The Romantic image of the artist as a rebellious outsider is powerful and enduring. As Stahl points out, however, "recording artists also typically work under unequal contracts and must hand over long-term control of the songs and albums they produce to their record companies." A less Romantic perspective on musicians would see them as carrying out "act[s] of labour within the industrialised process of cultural production", even if cultural work may seem "hardly like work at all" (Banks, 2007, p. 3–4). Typically, musicians' contracts with a record label are exclusive, meaning that the artist cannot record for anyone else without permission, and assignable, meaning that they can be bought and sold without consulting the artist. Finally, as Stahl points out, recording contracts typically cover not a fixed period of time but a number of album options – which can close off access to a competitive market for periods of 20 years or more. Far from the lone genius of Romantic thought, we can conceive of artists as carrying out creative labour – not always in enviable conditions. As Stahl argues, then:

> The recording artist – the successful recording artist, in particular – is a double figure. On the one hand, she is a symbolic figure offered for our consumption, contemplation, and identification; she enacts forms of expression, autonomy, and desirability, seeming to encapsulate some of our society's most cherished virtues and values. On the other hand, she is a political and economic actor, a working person whose contractually governed relationship to her company is sometimes one of real subordination. In this doubleness, the recording artist embodies a paradox: as an agent of self-expression under contract to a major entertainment conglomerate or a subsidiary company, the recording artist is both autonomous and the target of control. He [sic] must be free to generate new material and unfree when it comes to the labour and intellectual property covered by the contract. (Stahl, 2013, p. 2–3)

Artists' relationships with their record labels may have been at times subordinate, but they were at least contractually defined. Relationships between artists and streaming platforms, by contrast, are much more vague.

Music streaming algorithms: beyond the black box

Since they are both powerful and mysterious, the algorithms deployed by music streaming services might seem to illustrate an information asymmetry that is widespread in the information economy: 'corporate actors have unprecedented knowledge of our daily lives, while we know little to nothing about how they use this knowledge' (Pasquale, 2015, p. 9). The contemporary world, for Pasquale, resembles a one-way mirror or black box. Recommendation engines, for instance those operated by Amazon and YouTube, influence the choices we make – and that influence may not be benign. 'The economic, political, and cultural agendas behind their suggestions are hard to unravel. As middle-men, they specialise in shifting alliances, sometimes advancing the interests of customers, sometimes suppliers: all to orchestrate an online world that maximises their own profits' (Pasquale, 2015, p. 9). This information asymmetry, characteristic of what Srnicek (2016) calls "platform capitalism", is now fairly well established, acknowledged not only by scholars but in popular literature (O'Neil, 2016). Also increasingly established – and again noted by O'Neil (2016) – is the understanding that claims to impartiality by the corporate actors that deploy algorithms should be received with a degree of scepticism. As Gillespie (2014, p. 180) states, "the performance of *algorithmic objectivity* has become fundamental to the maintenance of these tools as legitimate brokers of relevant knowl-edge" (emphasis in original). Yet the impression that algorithms are neutral and objective is, for Gillespie (2014, p. 179) as for many others, "a carefully crafted fiction."

These twin insights – firstly, the acknowledgement of a fundamental information asym-metry, and secondly, an understanding of algorithms as not neutral or impartial – are important in analysing music streaming algorithms, which are, to some extent, black boxes. Yet the black box analogy can create the impression that, if we were only able to peer inside one of these black boxes, we would discover a secret key that would somehow "explain" or "solve" the mysteries of music streaming. Our approach in this article is instead informed by the more novel suggestion, proposed by Gillespie (2014, p. 169), that we should not attempt to understand algorithms in isolation; instead, we must "unpack the warm human and institu-tional choices that lie behind these cold mechanisms". This argument has been convincingly developed by Bucher (2018), who has made a strong case for moving beyond the black box analogy to a consideration of settings and contexts – a shift, in other words, "from what algorithms are to what they do" (Bucher, 2018, p. 42). Bucher's contribution is to argue that examining algorithmic power does not require an examination of algorithms themselves. Algorithms, Bucher argues "do not work on their own but need to be understood as part of a much wider network of relations and practices" (p. 20); 'we must not lose sight of the "human decision-making processes and programming that precede any algorithmic opera-tion" (p. 35). The black box analogy is insufficient, then, because algorithmic systems themselves are simply one part of a larger picture. "Algorithms are socio-material practices, not merely a set of coded instructions" (Bucher, 2018, p. 152); they are not neutral but, instead, "reflect the values and cultural assumptions of the people who write them" (p. 90). Netflix or YouTube deploy "not one algorithm but a collection of algorithms, working together to create a unified experience" (Bucher, 2018, p. 47), and contemporary music streaming services likewise combine many different algorithms to control the various elements to what the user sees on their screens. Algorithms, in other words, are multiple. Algorithms also exist in a state of flux: online platforms are typically engaged in constant A/B testing, as those *House of Cards* trailers illustrate. Users are essentially participating in giant focus groups even if they are not aware of it, resulting in 'the logic of "constant change" (Bucher, 2018, p. 48). Algorithms are not static or stable; instead, they can be "easily, instantly, radically and invisibly changed" (Gillespie, 2014, p. 178). The experience of any given platform will vary user to user; "the culture of experimentation complicates any effort to know algorithms, as the question inevitably arises as to which version, what test group, or what timeframe we are talking about" (Bucher, 2018, p. 48). Algorithms are not permanent but eventful, forever in a process of becoming (Bucher, 2018, p. 49).

In considering the algorithms utilised by music streaming services, then, we need to do more than imagine what might be hidden within the black boxes. Opacity is by no means limited to the algorithms. Streaming deals, too, are opaque: there is "much confusion" as to "how, exactly, streaming services are being licensed, how it is calculated what digital service providers (DSPs) must pay, and how that money is then processed and shared by the music rights industry" (Cooke, 2018, p. 8). Cooke (2018) identifies various reasons for this confusion: the complexity of streaming deals; the fact that the record industry (concerned with recordings) and the music publishing sector (concerned with songs and compositions) do not always license in the same way; variation between countries; the fact that most streaming deals are essentially revenue share agreements, making payments per-use difficult to predict; the fact that the specifics of many music streaming deals are kept secret due to non-disclosure agreements; and, finally, the fact that those responsible for developing new licensing arrangements have not always been effective in communicating these arrangements to the rest of the industry. Each month, as Cooke explains, a typical streaming service will pay a cut of its revenue to record labels and music publishers, calculated as a proportion of overall consumption; the copyrights in songs are treated by the record industry as distinct from the copyrights in sound recordings, with the former typically dealt with by music publishers, while the latter are typically controlled by record labels. Typically, Cooke suggests, a record label might expect to receive 50–60% of revenue; a music publisher might receive 10–15%; and the streaming service might retain approximately 30%. Yet "every deal is different, and usually secret" (Cooke, 2018, p. 15). Simply seeing inside the algorithmic back box will not shed much light on streaming platforms if streaming deals remain behind closed doors.

Also opaque is the extent to which streaming, while clearly benefiting the record industry in general, is benefitting individual artists. Streaming services tend to cite top-level statistics: Spotify, for instance, states that it has paid $8 billion to rights holders since its launch (IFPI 2018). What is less clear, however, is how much of that money reaches artists and songwriters. Even established artists such as Thom Yorke, Taylor Swift and David Byrne have complained about relatively low payments from streaming platforms, and the extent to which revenue from the sale of Spotify shares is passed from labels to artists remains a controversial topic (Paine, 2018). If anything, less commercially success-ful artists have more to complain about, since what money does reach artists and song-writers tends to go to established stars. Spotify founder Daniel Ek has suggested that only 0.733% of acts on Spotify are in the "top tier" in revenue terms – a tier that enjoys "material success" and the ability to "live off their work" (Ingham, 2018). More than 99% of audio streaming, it is claimed, is of the top 10% of tracks (Krukowski, 2018). Mulligan (2018b) has gone as far as to suggest that we may be witnessing the "end of the break-through artist", pointing out that not one of America's ten top-selling albums in 2017 was a debut; there was only one debut in the UK's top ten in the same year. "Just 30% of Spotify's most streamed artists in 2017," Mulligan (2018b) continues, "released their first album in the prior five years." While streaming is often presented as more democratic than traditional radio, then, there are reasons to treat the claim with caution. And again, the issue is much bigger than an algorithmic black box. Cooke (2018, p. 77) suggests that, on the sound recording side, a label might take perhaps 46.4% of revenue from relevant streams, with 11.6% going to the artist (assuming a total allocation of 58% to the *recording*); on the composition side, a publisher might take 3.6% of income from relevant streams and a songwriter 8.4% (assuming a total allocation of 12% to the *song*). This leaves the streaming platform with 30%. Yet as Cooke (2018, p. 76) acknowledges: "Quite how money is shared varies according to: each deal between a DSP and a rights owner; each artist and songwriter's individual label and publishing contracts; and collecting society conventions. Splits are also evolving because DSP deals are renegotiated every few years and revenue share arrangements have been altered slightly." There is, then, not one revenue share split but many – and, like the algorithms, these splits are forever in flux.

There are other reasons, too, why we need to look beyond algorithmic black boxes for a proper understanding of music streaming. A selling point of Apple Music, for instance, is its use of human, rather than algorithmic, recommendation engines, and many of Spotify's highly successful playlists, such as RapCaviar are also human-driven. Consider the high-profile hirings by Apple Music of BBC Radio 1 presenter Zane Lowe in 2015, and Charlie Sloth, a Radio 1 and 1Xtra presenter, in 2018. Consider, too the fact that DJ Semtex has recently left BBC 1Xtra, in part to increase his commitment to his Spotify podcast. Similarly high-profile hirings have been made at management level: former Radio 1 Head of Music George Ergatoudis is now the UK head of Apple Music, following a stint at Spotify; Austin Daboh, former Head of Music at 1Xtra, is now Head of Shows and Editorial, UK at Spotify; 1Xtra editor Ryan Newman also moved to Apple Music in 2018. To focus entirely on the algorithms used by these services, then, is to miss the bigger picture. Streaming platforms have many controlling inputs, of which computer code is just one. The system is not static, but continually developing through changes to the code, and interaction (by artists and listeners) with the service and the catalogue. Some Spotify playlists are algorithmically generated, while others, such as RapCaviar and Today's Top Hits, are entirely curated by humans – albeit humans likely to be examining data when making curatorial decisions. Artists and their representatives have always had to work to attract the attention of gatekeepers and intermediaries, and they now have to be cognisant of the control structures of streaming services to successfully reach an audience.

The inverted panopticon: a continuum of (in)visibility

Online platforms are frequently discussed in terms of surveillance capitalism (Bellamy Foster & McChesney, 2014; Fuchs, 2012; Zuboff, 2015), a model sometimes compared to Foucault's panopticon (1984, p. 206). Yet Bucher (2018, p. 84) insightfully inverts Foucault's model. The threat of the panopticon, after all, was that of permanent visibility. With algorithms, Bucher argues, we are dealing not with a "threat of visibility" but a "threat of invisibility": "the problem is not the possibility of being constantly observed but the possibility of constantly disappearing, of not being considered important enough" (Bucher, 2018, p. 84). This threat of invisibility is of clear relevance to artists. Given the importance of streaming as a revenue source, the most obvious threat in this regard would be an outright ban, removing the possibility of earning revenue from streams (not to mention the other revenue streams, from live performance and direct-to-fan services, that are to a significant extent reliant on visibility on streaming services). According to Spotify, 2017), no music has ever been banned from their platform. The banned/not banned binary, however, may be too simplistic. Might streaming services make more subtle changes to the visibility of a given artist? Might we, for instance, be able to identify artists subject to a "shadow ban"? Within the world of online forums, a shadow ban is a process in which a user is banned from part of a service, for instance, from posting on a particular thread, in a manner that is invisible to that user. The phrase has also been applied to social media, with President Donald Trump, for instance, acccusing Twitter of shadow banning certain Republicans from its platform (Stack, 2018). Again, the crucial feature of a shadow ban is that the individual concerned is unaware that such a ban exists: 'When a person is shadow banned, their posts on a platform are rendered essentially invisible to everyone but themselves. Their experience using a site may not change – they feel like they are still posting normally – but other people cannot see the material they produce' (Stack, 2018).

The notion of a shadow ban suggests that the banned/not banned binary is, indeed, too reductive. In fact, we suggest that even the question of whether or not an artist is subject to a shadow bans is too blunt a tool for examining the range of subtle practices that may be occurring within music streaming services, not least because it suggests that platforms can only ban (shadow or otherwise) when, in fact, they can also *boost* visibility. Instead, we propose a sliding scale of possibilities, a continuum along which artists and their music could be both upgraded and downgraded. We can identify five possible points on this continuum. Firstly, a song or an artist could receive a *public upgrade*: an artist's album, for instance, could appear on the interface homepage. Secondly, a song or artist could receive its antithesis, a *public downgrade*: for example, an acknowledgement that an artist has been made less visible on a given platform. Thirdly, a song or artist could receive a *shadow upgrade*: a song, for instance, could become more likely to be selected within algorithmically-generated radio stations. Fourthly, a song or artist could be subject to the opposite, a *shadow downgrade*: for example, an artist might not appear clearly within search results, requiring a user to spend additional time finding their catalogue. Fifthly, a song or artist might receive an *outright ban*: in other words, that song or artist would not appear on a given service at all.

As discussed, we are not aware of any artists receiving an outright ban from a streaming service, but we now provide examples to illustrate the other four points on this continuum. First, public downgrade and upgrade. In May 2018, it was announced that Spotify had removed the music of R&B artist R Kelly from its playlists as part of a new "hate content and hurtful conduct" policy. Users were still able to find R Kelly's music if they searched for it, but Spotify would no longer actively promote his recordings through their own playlists and recommendations. The move was in response to the #MuteRKelly social media campaign, which called for the singer to be boycotted due to sexual assault allegations – allegations that R Kelly denies. The move was noteworthy, in part, because it raised the question of whether streaming services should be making moral judgements – not to mention the subsequent question of the grounds on which such judgements might legitimately be made. For the purposes of this article, however, the move was noteworthy for a different reason: Spotify's acknowledgement that they could, and would, tweak their algorithms to affect the visibility of a given artist. The R Kelly decision was fairly widely reported in the media, and Spotify issued a public statement on the subject. This was, in other words, a public downgrade. Under the same "hate content and hurtful content" policy, Spotify also removed from its playlists the music of rapper Jahseh Dwayne Onfroy, better known as XXXTentacion. At the time, the rapper was on trial for false imprisonment, witness tampering and aggravated battery of a pregnant woman, charges to which he pled not guilty. This too, then, was a public downgrade. Strikingly, XXXTentacion is also an example of a public upgrade on the streaming service. When the rapper was shot dead in June 2018, in an apparent attempted robbery,

Spotify then began to promote his music with the message "Rest in peace, XXXTentacion" shown on the homepage of many users, alongside a promoted playlist (Cush, 2018) – the platform, consequently, was accused of hypocrisy. The cases of R Kelly and XXXTentacion are particularly notable as public acknowledgement by Spotify that "downgrading" of artist visibility is not only technically possible but a practice in which they were willing to engage. Spotify's acknowledgement gives rise to further questions that we should ask not just of Spotify but of all streaming services. Are there other downgrades that have not been publicly acknowledged? On what grounds other than moral censure might a streaming service justify downgrading or upgrading artists in this fashion? It should be noted that, by the time XXXTentacion was shot dead, Spotify had dropped its "hate content and hurtful conduct" policy; the policy, indeed, barely lasted a fortnight. Founder Daniel Ek admitted that the streaming service "rolled this out wrong and we could have done a much better job", insisting that the platform did not intend to become a "moral police" (Wang, 2018). Furthermore, it was rumoured that Troy Carter, Spotify's Global Head of Creator Services, threatened to quit over the policy (Halperin & Aswad, 2018), and that representatives for artists including Kendrick Lamar called Ek to express their frustration (Sargent 2018) – itself an interesting hint of the power of the (high-profile) artist in a supposedly algorithmic age. The broader question of algorithmic accountability, however remains.

We move now to the shadow downgrades, the most rudimentary example of which can be experienced by undertaking the admittedly grim task of searching on Spotify for artists convicted of extremely serious and deeply repugnant sexual crimes. At least for the authors of this paper, for instance, neither Gary Glitter (convicted of possessing child pornography in 1999, then jailed in 2006 in Vietnam, and 2015 in the UK, on child sexual abuse charges) nor Lostprophets (whose singer, Ian Watkins, was sentenced to 35 years in prison for child sex offenses in 2013) appear in the results of an initial search for their respective names – although in both cases, the artists and their catalogues *do* appear when the "see all" button is clicked. This would appear to be a shadow downgrade in action. The music still exists on the platform, and so fits with Spotify's assurance that no music receives an outright ban. However, additional work is required to find the music. Not all high-profile criminal artists, however, appear to have been downgraded in this manner. The reasons for these disparities is not clear. Examples of the final category we identify, the shadow upgrade, are hardest to come by. Indeed, we have no evidence that such practice is occuring; we simply speculate as to what might be possible. While it is easy enough to see that Gary Glitter, for instance, has been made less visible in search results, it may be impossible to show that another artist is getting the opposite – i.e. preferential treatment. It would be entirely possible, however, for a shadow upgrade to make one particular artist with the first name of "James" the top search result rather than another, or to make songs by a particular artist more likely to be selected by a particular algorithm. However, this would also be almost impossible to evidence. The term "shadow" indicates that not only will the artist or individual user be unaware of a downgrade, but also that the criteria by which such a ban might be justified are also hidden. As previously described, there is no single algorithm or service, but many. There is therefore no means of appeal, as a downgrade is almost impossible to prove.

Since algorithms are both multiple and constantly in flux, the scenarios identified above should not be understood as mutually exclusive alternatives; it might be possible for an artist to be upgraded on one user's profile but simultaneously downgraded on another's.

This could, indeed, be precisely the sort of A/B testing in which a streaming service might might wish to engage. Netflix, Bucher (2018) states, might be running not only two variants of an algorithm, but five or ten, with multiple tests being carried out in parallel. As a result, we should think of Netflix as a work in progress: "So-called A/B testing is integral to the culture of experimentation that governs platforms as a means of assessing and comparing how well different versions of the algorithm perform," although these experiments are "invisible" to users (Bucher, 2018, p. 48). This is also true of Facebook: Bucher describes the platform's news feed as constantly being fine-tuned, "there is no clearcut way in which a certain state of the feed can be easily discerned" (Bucher, 2018, p. 81). The same is true of most streaming services. Due to the highly personalised nature of the systems, what one user sees could be wildly different from the next. Much as there is no single algorithm to study, so there is potentially no single system on which artists are downgraded or upgraded, but a dynamic set of systems, continually managed by both human and machine input. It is for this reason that the term "shadow ban" does not suffice. Artists may not be removed from all playlists, but from a certain proportion. That proportion of listeners are shown a slightly different version of a given webpage, and their behaviour is used as a point of comparison: which version gains more clicks or longer listens? Since A/B testing is invisible, neither the user, nor the artist, will ever be aware of these changes in design. Additionally, due to the use of machine learning, even a human engineer may not be aware of such changes – Discover Weekly is essentially a unique playlist for each and every Spotify user, and so no human is ever aware of all 100m+ playlists in circulation. Only the machine system has such awareness.

The public downgrades of R Kelly and XXXTentacion appear to have been made for moral reasons; that appears true of the shadow downgrades of Gary Glitter and Lostprophets too. However there are a multitude of other possible influences over upgrades and downgrades. It has been alleged, for instance, that Spotify is "burying" the music of artists who introduce new material exclusively with their rival Apple Music (Shaw & Santiano, 2016). Katy Perry has allegedly been a victim of this practice: her 2016 track "Rise", it is claimed, was "continually blackballed" on Spotify's biggest playlists, although it was not downgraded in search results (Ingham, 2016a). This "blackballing", it is alleged, was in response to the fact that the track was originally released as a "windowed" iTunes and Apple Music exclusive, not appearing on Spotify for a full week. To be clear, Spotify insisted that it did not bury Katy Perry or other artists close to Apple or Tidal in search results (Ingham, 2016a); but Ingham insists that the streaming service *did* keep them off its playlists – a similar situation to R Kelly and XXXTentacion in 2018, although one that was not publicly discussed as part of a PR campaign. Services have other business relationships to maintain that may somewhat influence internal decisions. For example since 2008, Spotify has reportedly had major investments from organisations including Goldman Sachs, the Coca Cola Company, and Dragoneer Investment Group (Ingham, 2016b), and back in 2015 they made €68m from advertising (Ingham, 2015). Financial investment through funding and advertising spend is important for the business both for growth, and to keep shareholders content. Artists themselves also may exert power and influence over the upgrades and downgrades, similar to the phone calls made to Daniel Ek around the XXXTentacion public downgrade. For example Bruno Mars (in November 2018, the 29th most streamed artist in the world on Spotify) played at Daniel Ek's wedding to Sofia Levander in 2016; and Ed Sheeran (13th most streamed artist on Spotify), has spoken publicly in support of Spotify (Associated Press, 2015; Dredge, 2014; Savage, 2014).

Even established, top-tier artists could be vulnerable. What if a service takes issue with an artist's political statements, or with a band performing in Israel? To be clear, this is not simply about algorithm design. Being downgraded by a particular algorithm is just one way power may be exerted over an artist – royalty rates could be changed, equity may be used as an incentive. The systems in place are influenced by many factors, only one of which is the structure of individual algorithms. However, financial pressure, personal relationships, human bias, market competition, and licensing agreements all guide music recommendations. While we do not know whether or not particular artists have been up or downgraded, this lack of knowledge is precisely the point. And looking inside the black box would not shed the required light.

Conclusion: implications for future research

With most income from recorded music now generated through streaming services, and fans finding music through tools – text search, categorical lists, playlists, hyperlinks – controlled exclusively by these services, understanding these services and tools is of primary importance to the contemporary recording artist, as well as to scholars of the music industries. The notion persists that there is, within a given streaming platform, a black box, to which artists and their managers must gain access if they wish to understand and effectively benefit from that streaming service. In truth, there is no such "secret sauce". Scholars and artists alike, then, need to look beyond specific algorithms, and to understand that the power behind recommendation systems is primarily located elsewhere. This power could conceivably result in downgrades or upgrades, be they "public" or "shadow", each with potentially significant effects on artists who are increasingly reliant on streaming services as a source of income from recorded music, as well as a means to gain the profile required to earn meaningful income from other revenue streams including live performance and direct-to-fan services. There was some excitement around the launch of Spotify for Artists, which allowed artists to access detailed data on their listeners, as well as to submit music to playlist curators, and this is indeed a positive step for creators. Yet to focus only on algorithms, or on data, is to neglect the fact that any streaming service will be subject to a number of influences: from employees, shareholders, advertisers, record labels, music publishers and so on; from rivalries with other streaming services; from public opinion, which may result in moral censure of particular artists. Any and all of these may have an impact on recommendation systems. An increased understanding of the subtleties of downgrades and upgrades, then, is necessary to truly understand the power of streaming services, and the potential financial consequences on artists. Downgrades and upgrades do not simply highlight the design of a given recommendation algorithm, but reflect the power that streaming services have over artists and their income. Scholars and researchers need to consider the entire system – not only the mythical standalone algorithm – and be mindful that this system is constantly becoming, constantly in flux, and therefore never fully visible. To obsess over an algorithmic black box at this or that streaming service is to neglect the fact that power dynamics extend far beyond code. Further research to shed light on power and control within music streaming services is therefore required.

Disclosure statement

No potential conflict of interest was reported by the authors.

References

Airoldi, A., Beraldo, D., & Gandini, A. (2016). Follow the algorithm: An exploratory investigation of music on YouTube. *Poetics*, *57*, 1–13. doi:10.1016/j.poetic.2016.05.001

Associated Press. (2015, October 22). Ed Sheeran talks hangovers and Spotify. *Billboard*. Retrieved from https://www.billboard.com/articles/columns/pop-shop/6738253/ed-sheeran-hangovers-spotify

Aswad, J. (2018, November 1). Spotify Reaches 87 Million paid subscribers. *Variety*. Retrieved from https://variety.com/2018/music/news/spotify-reaches-87-million-paid-subscribers-1203017004

Banks, M. (2007). *The politics of cultural work. Houndmills*. Basingstoke, Hampshire, UK: Palgrave Macmillan.

Beer, D. (2017). The social power of algorithms. *Information, Communication & Society*, *20*(1), 1–13. doi:10.1080/1369118X.2016.1216147

Bellamy Foster, J., & McChesney, R. (2014, July 1). Surveillance capitalism. *Monthly Review*. Retrieved from https://monthlyreview.org/2014/07/01/surveillance-capitalism

Bucher, T. (2018). *If … then: Algorithmic power and politics*. Oxford, UK: Oxford University Press.

Chang, L. (2018, May 15). Apple music hits 50 million users, doubles down on original content. *Digital Trends*. Retrieved from https://www.digitaltrends.com/music/apple-music-50-million-users

Cooke, C. (2018). *Dissecting the digital dollar* (2nd ed.). London, UK: Music Managers Forum.

Cush, A. (2018, June 19). Spotify has no idea what it's doing with XXXTentacion. *Spin*. Retrieved from https://www.spin.com/2018/06/spotify-xxxtentacion-playlists-death

Dredge, S. (2014, September 30). Ed Sheeran talks Spotify royalties. *The Guardian*. Retrieved from https://www.theguardian.com/technology/2014/sep/30/ed-sheeran-spotify-streaming

Foucault, M. (1984). *The Foucault Reader: Discipline and Punish, Panopticism*. New York, NY: Penguin Books.

Fuchs, C. (2012). The political economy of privacy on Facebook. *Television & New Media*, *13*(2), 139–159. doi:10.1177/1527476411415699

Gillespie, T. (2014). The relevance of algorithms. In T. Gillespie, P. J. Boczkowski, & K. A. Foo (Eds.), *Media technologies: Essays on communication, materiality, and society* (pp. 167–194). Cambridge, Mass: MIT.

Halperin, S., & Aswad, J. (2018, May 14). Is troy carter out at spotify? *Variety*. Retrieved from https://variety.com/2018/music/news/is-troy-carter-out-spotify-1202808085

IFPI. (2018, April). IFPI Global Music Report 2018. IFPI. Retrieved from https://www.ifpi.org/news/IFPI-GLOBAL-MUSIC-REPORT-2018

Ingham, T. (2015, April 19). Spotify ad revenue jumped 53% in Q1. *Music Business Worldwide*. Retrieved from https://www.musicbusinessworldwide.com/spotify-ad-revenue-jumped-53-q1

Ingham, T. (2016a, August 29). Katy Perry knows exactly how much Spotify is 'punishing' Apple exclusive artists …. *Music Business Worldwide*. Retrieved from https://www.musicbusinessworldwide.com/katy-perry-knows-exactly-how-much-spotify-is-punishing-apple-exclusive-artists

Ingham, T. (2016b, November 18). Spotify co-founders own an estimated $1.8bn stake in company. *Music Business Worldwide*. Retrieved from https://www.musicbusinessworldwide.com/spotify-co-founders-own-1-8bn-stake-in-company-say-estimates

Ingham, T. (2018, March 25). The odds of an artist becoming a 'top tier' earner on Spotify today? Less than 1%. *Music Business Worldwide*. Retrieved from https://www.musicbusinessworldwide.com/the-odds-of-an-artist-becoming-a-top-tier-earner-on-spotify-today-less-than-1

Krukowski, D. (2018, January 30). How to be a responsible music fan in the age of streaming. *Pitchfork*. Retrieved from https://pitchfork.com/features/oped/how-to-be-a-responsible-music-fan-in-the-age-of-streaming

Moon, A., & Nellis, S. (2018, October 17). Spotify takes minor stake in music distributor distrokid. *Reuters.* Retrieved from https://www.reuters.com/article/us-spotify-tech-distrokid/spotify-takes-minor-stake-in-music-distributor-distrokid-idUSKCN1MR325

Mulligan, M. (2018a). *Emerging music markets: Streaming's third wave.* Retrieved from https://musicindustryblog.wordpress.com/2018/08/10/emerging-music-markets-streamings-third-wave/

Mulligan, M. (2018b). *How streaming is changing the shape of music itself (Part 1).* Retrieved from https://musicindustryblog.wordpress.com/2018/09/06/how-streaming-is-changing-the-shape-of-music-itself-part-i/

O'Neil, C. (2016). *Weapons of math destruction: How big data increase inequality and threatens democracy.* London, UK: Penguin.

Paine, A. (2018). Warner music group letter reveals Spotify windfall terms for artists. *Music Week.* Retrieved from http://www.musicweek.com/labels/read/warner-music-group-letter-reveals-spotify-windfall-terms-for-artists/073573

Pasquale, F. (2015). *Black box society: The secret algorithms that control money and information.* Cambridge, Mass: Harvard University Press.

Prey, R. (2017). Nothing personal: Algorithmic individualisation on music streaming platforms. *Media, Culture and Society, 40*(7), 1086–1100. doi:10.1177/0163443717745147

RIAA. (2018, September). 2018 Mid-year shipment & revenue statistics. *RIAA.* Retrieved from https://www.riaa.com/reports/2018-mid-year-shipment-revenue-statistics-riaa

Rogers, J. (2013). *The life and death of the music industry in the digital age.* London, UK: Bloomsbury.

Sargent, J. (2018). Did Kendrick Lamar threaten to pull his music from spotify over XXXTentacion Ban? SPIN. Retrieved from https://www.spin.com/2018/05/kendrick-lamar-spotify-xxxtentacion-report

Savage, M. (2014, December 12). Ed Sheeran 'owes career to Spotify'. *BBC.* Retrieved from https://www.bbc.co.uk/news/entertainment-arts-30436855

Shaw, L., & Santiano, A. (2016, August 26). Spotify is burying musicians for their Apple deals. *Bloomberg.* Retrieved from https://www.bloomberg.com/news/articles/2016-08-26/spotify-said-to-retaliate-against-artists-with-apple-exclusives

Smith, M. D., & Telang, R. (2017). *Streaming, sharing stealing: Big data and the future of entertainment.* Cambridge, Mass: MIT.

Spotify. (2017, October 30). [Music] block/hide/blacklist tracks or artists. *Spotify Community.* Retrieved from https://community.spotify.com/t5/Closed-Ideas/Music-Block-Hide-Blacklist-Tracks-or-Artists/idi-p/22389

Srnicek, N. (2016). *Platform capitalism.* Cambridge, UK: Polity.

Stack, L. (2018, July 26), What is a 'shadow ban,' and is Twitter doing it to republican accounts? *The New York Times.* Retrieved from https://www.nytimes.com/2018/07/26/us/politics/twitter-shadowbanning.html

Stahl, M. (2013). *Unfree masters: Recording artists and the politics of work.* Durham, UK: Duke University Press.

Wang, A. (2018, May 31). Spotify admits its R. Kelly Ban was 'rolled out wrong'. *Rolling Stone.* Retrieved from https://www.rollingstone.com/music/music-news/spotify-admits-its-r-kelly-ban-was-rolled-out-wrong-627739

Wikström, P. (2013). *The music industry: Music in the cloud* (2nd ed.). Cambridge, UK: Polity.

Zuboff, S. (2015). Big other: Surveillance capitalism and the prospects of an information civilization. *Journal of Information Technology, 30,* 75–89. doi:10.1057/jit.2015.5

Revenue, access, and engagement via the in-house curated Spotify playlist in Australia

Benjamin A. Morgan ⓘ

ABSTRACT

This article analyzes the perception of recording industry change in Australia through a focus on the curated playlist as a sociotechnical intermediary. During the rapid expansion of commercial music streaming services between 2016 and 2018, musician and recording industry enthusiasm for revenue growth was tempered by trepidation surrounding change. This article presents findings from interview research conducted with Australian artists and music industry professionals to provide insight on the promotion of popular music within the often opaque blend of human and algorithmic structures inside the digital music streaming commodity. It explores the apprehension and enthusiasm over the affordances of the new recommendation and discovery ecosystems, centered around the in-house curated Spotify playlist. Debates surrounding the topics of revenue, access, and engagement are explored. Analysis will demonstrate the increasingly influential role music streaming services and recommendation tools are perceived to play in the sectors of retail, promotion, and distribution.

Introduction

Between 2016 and 2018, popular press coverage of music streaming services kept up the tradition of sensationalist rhetoric that has surrounded the emergence of new technologies which effect popular music. Reporting frequently relied on narratives portraying music streaming as a savior resurrecting the recording industry from the dead (Ellis-Petersen, 2016; Nicolaou, 2017). These coexisted with skeptical reports claiming that streaming was unfair to artists, or "how the music industry is putting itself out of business" (Gerber, 2017). As scholars of media and/or technology should expect, the discussion around streaming services within the industry was far more nuanced and specific than the blunt moralistic frames presented in the popular press. Those meta-narratives and moral frames around streaming are simplistic distractions from more useful questions regarding what new practices, concerns, and norms emerge when the affordances of new technologies affect the popular music commodity.

Acknowledging that the recording industry of the 20th and early 21st centuries has so far been a superstar model where a handful of sustainable successes fund failures (Marshall, 2013) and that "the majority product of the music industry is not success but failure" (Jones, 2003, p. 148), my research is based on the belief that there are "more and less equitable ways to divide the industry's wealth, and to develop methods to insure its continuing growth and innovation" (Sinnreich, 2016, p. 154). Conducted as the qualitative element of Australian Research Council linkage project *Music usage metrics and the future of the Australian music industry*, it examines the role of streaming technology in changing practices through interviews with Australian industry stakeholders and artists.

Given that there are both empowering and exploitative narratives around the affordances and structures of streaming to be explored, this study of how stakeholders view change in the Australian music industries[1] contributes to critical media industries research, which "needs to be underpinned by an engagement with social theory, with empirical work (which means an engagement with methodological problems and dilemmas), and with a politics opposed to inequality and injustice" (Hesmondhalgh, 2009, p. 253). This qualitative project on the production and distribution of popular music in Australia is intended to inform existing and future systems for the generation of revenue from popular music with evidence reflecting the needs and concerns of the Australian recording industry as streaming was surpassing other formats as the primary revenue source for recordings.

Music streaming services were seen to be impacting industrial practice and strategy in a variety of ways, but the scope of this article is to explore stakeholder views in the Australian recording sector around the Spotify in-house curated playlist, in order to help make sense of related debates regarding revenue, access, and engagement. It will examine the discourse around changing practice within both the realms of commercial marketing and promotion strategy (*distribution*). As a micro study on media production and distribution, it seeks a better understanding of the "intertwining of economics and culture in capitalism" while acknowledging that there are contradictory elements of both emancipation and domination to be found in debates around media industry production under capitalism (Hesmondhalgh, 2009, p. 253). While labor practice and commercial strategy are often the discussion topics, this research is ultimately concerned with the meanings and standards constructed through practice.

Interview sample and Australian context

My research follows the attitudes of artists and professionals in the Australian music industries around technology in semi-structured interviews. The interview sample (n = 30) consists of emerging and aspiring popular musical performers (n = 7)[2] and professional stakeholders involved in the monetization of artist brands across multiple industry sector roles: large independent record labels (n = 2), small independent record labels (n = 2), artist managers (n = 5), publishers (n = 4), aggregators/distributors (n = 2), live concerts (n = 2), marketing services (n = 2), a retailer (n = 1), a technology services consultant (n = 1), royalties accountant (n = 1), and a radio producer (n = 1).

The Australian popular music sector has been described as a "small marketplace, heavily reliant on cultural imports" where it "can be a difficult place to succeed and to create ongoing careers," due largely to a low population density and large distances between major cities (Strong & Rogers, 2016). Despite its geographic isolation and perceived fondness for importing popular music, it is also positioned by industry as a strong exporter hoping to be the source of "five per cent market share of the global music market by 2030" (Australia, 2019, p. iv), while domestic consumption of recordings was ranked as the eighth largest market in the world overall in 2018 (IFPI, 2019, p. 13).

The risk-reducing economic structures of stardom, which can be found wherever the global recording industry is well-established, encourages overproduction and is a model where very few artist brands are able to become sustainable successes (Marshall, 2013; Negus, 1999). However, there is an Australian angle to mediating this prevalence of failure. Based on research conducted with musicians in Brisbane, (Rogers, 2013) argues that the production of popular music in Australia is centered around amateurs and hobbyists, with professionalization into a sustainable career often seen as secondary to the immediate pleasures of producing. This suggestion that *success* is not primarily defined as a sustainable career by many Australian musicians is also supported by Hughes, Keith, Morrow, Evans, & Crowdy (2013), who found that intermediate success and satisfaction in achievement were more appropriate descriptors for success in Australia than traditional metrics of financial independence or peer respect/recognition. All artists that were interviewed for this study relied on income derived outside of music production. All seven had hired staff to work on campaigns and had released records. Four employed managers, and five employed booking agents at the time of interviews. While this sample lacks the voice of any stars deriving their entire livelihood from popular music production, it is reflective of the status of the vast majority of Australian artists in the process of pursuing careers as producers of popular music.

The Australian music industries are supported through policies designed to protect and support the production of local content. Funded grants to support different stages of production and promotion for artist brands are offered through Federal organizations such as the Australia Council for the Arts and Sounds Australia, state bodies like Creative Victoria, Create NSW, and QMusic, as well as local and industry initiatives. One particular notable institution is the publicly-funded national youth broadcaster Triple J, "a unique blend of public service and niche marketing" (Albury, 1999, p. 55) widely regarded as the most important and influential gatekeeper for the youth music market. While policy to support local content within the streaming format was under discussion, interviews focused on how new technology affected everyday practice and strategy for promoting the artist brand within the existing promotional infrastructure of intermediaries and technological tools. Noting concerns raised by several scholars about applying research with creative workers in highly developed markets to the rest of the world (Alacovska & Gill, 2019; De Beukelaer, 2015; Morgan, 2019), the findings of this paper should be linked to Australian cultural market conditions.

I spoke with a broad range of sector professionals, and several participants held multiple roles. One interview ought not be interpreted as representative of an entire sub-sector role. All industry participants were either sole proprietors or held management positions at firms. Staff at multi-national major record labels/publishers and commercial streaming companies declined to participate in this research. It is important to note that audiences also play a central role in shaping practice, but this study is concerned with how industry stakeholders and artists contribute to the formation of standards and meaning around popular music. While several other studies have looked at how users engage specifically with streaming platform playlists (Hagen, 2015) or algorithmic recommendation systems (Karakayali, Kostem, & Galip, 2017) in consumption practice, this article examines the perspective of producers and professionals around promotion and distribution.

The interview methods and theoretical grounding are based on other scholarly studies on how music industries have reacted to digitalization and new tools (Baym, 2011, 2018; Hesmondhalgh & Meier, 2018; Meier, 2017; Nordgård, 2018; Wikström, 2013). The conceptual approach of Keith Negus (1999), who interviewed and observed music producers and professionals in order to analyze how their culture produces the industry, guides this investigation.

Locating the in-house curated spotify playlist: a new form of sociotechnical cultural intermediary

The intended object of interrogation, the in-house curated Spotify playlist is a specific entity within the digital music commodity known as a streaming service, a relatively new format ecosystem for consuming popular music. Streaming services offer various methods of algorithmic recommendation, search, and customization intended to keep the user within their ecosystem by continually suggesting more songs to listen to. The branded virtual space itself is the commodity on offer, and the commodity is defined no longer merely by the music being listened to, but by infrastructures which are intentionally designed to trap the user and keep them listening as long as possible by continually suggesting new songs (Seaver, 2018). This combination of distribution and discovery roles of the digital music streaming commodity has created new tensions arising from the expectation of these services to constantly provide promotion of songs in order to prolong consumption (Kjus, 2016).

The revenue-generating streaming services which monetize streams and pay copyright owners through licensing agreements were frequently referred to by participants as *DSPs* (for digital service providers), but also as online stores, subscription services, streaming sites, and platforms.[3] During this period in Australia, DSPs referred to Spotify, Apple Music, Google Play, Amazon, and Tidal. *DSP* appears to be a relatively new music industry term.[4] Stakeholders at larger music companies were fond of using it, but it was still unfamiliar to several participants. DSPs grant access to large catalogs owned by record labels, offer playlists and recommendations to promote recordings, and monetize the streams.

The recommendation features within these branded virtual spaces are new forms of *cultural intermediaries*. Originally used by Pierre Bourdieu to refer to "producers of cultural programmes on TV and radio or the critics of 'quality' newspapers and magazines and all the writer-journalists and journalist-writers" (Bourdieu, 1984, p. 325), this is a theoretical model for studying people (or in this case, an opaque hybrid of people and algorithms) who function as gatekeepers which acknowledges that they are involved in the shaping of popular taste. Bourdieu's concepts of non-economic capital are useful when discussing playlist pitching to make sense of the professional networks (*social capital*) and experience, education, and knowledge within the field (*cultural capital*) that are involved with gatekeeping and access (Bourdieu, 1986).

This concept has been useful for understanding the role of record labels, publicists, concert promoters, and many other cultural industry workers in adding value and contributing to the success or failure of popular songs and artist brands over the years (Meier,

2013; Negus, 1999, 2002). Powers (2015) conceptualization of *intermediation* moves this conception away from individuals towards a "dynamic process of circulation that involves people, symbolic forms, and objects as much as it involves the modes of transportation and transmission that allow these elements to be linked" (p.122). Morris (2015) builds on this by conceiving of DSPs as digital *infomediaries* which mine, monitor, and mediate their audiences, while noting that it is now particularly important to understand how algorithms are reconfiguring the work of cultural intermediaries. Webster, Gibbins, Halford, and Hracs (2016) combine Bourdieu's concept of cultural intermediaries with Actor-Network Theory's insistence on the relational ontology of human and nonhuman actors, arguing that the music recommender systems within the DSPs should be approached as a new form of "sociotechnical cultural intermediary" (p.141). This term is particularly appropriate given the combination of human and algorithmic curation involved in the management of Spotify's in-house playlists.

While Spotify is functioning as an *infomediary* at the institutional level, it was their recommendation ecosystem which had become the main topic of interest for artist managers, labels, and publishers interested in promoting releases by their artist brands. This recommendation ecosystem included featuring artist brands or recordings on the interface, algorithmic auto-generated tailored song suggestions based on past activity, and a large number of playlists. These different methods of recommendation are all gatekeeping mechanisms which allow certain recordings to be highlighted, suggested, or even automatically played in lieu of others. While all users and third-party institutions could create their own playlists and allow others to access them, it was the playlists which Spotify created and managed in-house which were perceived by participants to be the key to the largest potential audience. As I also learned, the degree to which algorithmic vs. human curation played a role in managing these playlists was not well-understood by participants.[5]

Curated playlist: the instrument of new affordances

The curated playlist is a mechanism whereby DSPs are able to combine format affordances in promotion, distribution, and retail of recordings which can aid in the discovery[6] of new songs, deliver them, and facilitate payment. These playlists are often based on genre, though Spotify's specialties include mood- and activity-based playlists that promise the ideal soundtrack to match an activity or a desired feeling, such as Boozy Brunch, Songs to Sing in the Car, Putting Baby to Sleep, or Deep Focus. While some of the playlists are focused on the concept of a musical theme (e.g. I Love My '00s R&B, JazzRap), the popularity of soundtracking-your-life playlists does indicate a shift towards a more contextual application of music based on the listener's location, activity, and state of mind. These particular playlists are examples of what some participants refer to as *lean back music* – a more passive approach to song selection that is often based on context.

While the exact procedures of how Spotify playlists are curated is opaque, journalists have profiled some of the human curators and their methods (Ugwu, 2016). Interview data confirms that Spotify, Apple Music, and Google Play employed human curators who accepted pitches to support new releases for playlists (and other online editorial features within the sites). These in-house curated playlists are different than user-generated personal playlists and those created by outside firms which are also known as third-party playlists. There is a market for promoting songs to the third-party playlisters, which do not enjoy the same placement and focus within the site. While some labels had looked into third-party playlisting promotion, none had hired them. Unless otherwise noted, curated in-house playlists are the type of playlist under discussion.

Playlists are the embodiment of the new affordances of DSPs – the ability to promote the discovery of a particular song, enable the delivery, charge the listener, and compensate the copyright owner(s). This consolidates formerly distributed power with one entity, which bypasses existing intermediaries. It also reduces cost and system friction and presents an alternate promotional pathway through which to promote recordings to radio broadcasting, the traditional gatekeeper of popular success. While the emergence of a new promotional pathway was welcomed, three areas where debate about playlists existed were around the topics of revenue, access, and engagement.

Revenue on playlists: enthusiasm and pessimism

The controversy surrounding artist royalties for streams was a high-profile debate well-covered in the popular media and analyzed by scholars (Marshall, 2015; Sinnreich, 2016). I anticipated that revenue rates would be a major topic. My first surprise finding was that artists and industry stakeholders were not interested in talking about streaming rates. Global and local journalists were continuing to report on criticism about transparency and fairness in streaming rates first raised by popular artists such as David Byrne (Byrne, 2015), Taylor Swift (Swift, 2015), or Thom Yorke (Dredge, 2013) and enforced by research white papers (Fair Music: Transparency and payment flows in the music industry, 2015; Lalonde, 2014). A major finding of this research is that participants were not (or were no longer) focused on rates of payment or transparency of payment by streaming sites as issues of concern.

Less-experienced performers were pessimistic about ever earning an income from streaming, e.g. "artists don't make any money from streaming, especially local artists" (Amy Wilson, personal communication, 2017). These skeptical statements came from the artists with lower stream numbers, and this skepticism about streaming revenue ever being substantial was also voiced by the smaller record label staff, who made more revenue from vinyl and still considered physical copies their focus. Enthusiasm about the growth of streaming revenue was strong with large labels, publishers, and managers, and perhaps most surprisingly, from the manager of unsigned Melbourne artist LANKS, who had experienced playlist and subsequent national radio support in 2016–17 without signing to a record label. I spoke with his manager, Lou Cuming.

> The opportunity that Spotify especially has given us with playlisting has meant that our income's grown from nothing to tens of thousands of dollars per year. So, purely from a discovery and a financial point of view, Spotify kind of represents the most lucrative and the most beneficial of the streaming services for what we're doing (Lou Cuming, personal communication, 2017).

Cuming's comments were the most enthusiastic that I encountered around streaming revenue in the interviews. The support LANKS was enjoying was indicative of a successful emerging act, which is important to differentiate from other aspiring unsigned acts who were not receiving the same level of playlist support. Of the seven artists interviewed, three had performed on songs which had stream counts in the millions on Spotify. None of those artists with higher stream counts verbalized the sentiment that the payments they were receiving were unfair.

The artists with lower stream counts and the smaller record label stakeholders gave consistent statements indicating the purchase of a physical copy of a recording or other merchandise was still the proper way to ensure financial support for a young artist. In contrast, the larger labels, publishers, and managers were all enthusiastic about the ability of DSPs to generate revenue, which industry reports would subsequently confirm were growing in Australia at very impressive rates. Reported streaming revenues accounted for 38% of the Australian recorded music market in 2016, growing to 54% in 2017, and climbing to 71% of a sector worth $526 million AUD in 2018 (ARIA, 2019; Australia, 2019, p. 6).

The narratives portraying streaming royalties as unfair to artists have been noted by Marshall (2015). The example of LANKS finding success and revenue without a record label via the playlist demonstrates that emancipatory claims of streaming enabling creators to bypass powerful intermediaries such as record labels can cite certain cases as evidence. It was not surprising that the artists with lower stream counts were skeptical of the potential of streaming revenue, while stakeholders who worked with portfolios of recordings could more easily see the impact on income. Overall, this interview sample was not critical of payment rates for streams, an unsigned artist's manager was very positive about the revenue, industry professionals were enthusiastic about the ability of playlists to generate revenue, and everyone viewed playlists as important promotional tools, even when they were skeptical about the revenue ever becoming significant for them.

Access to promotion through playlists: clear pitching and opaque placement

"Spotify's mixes are replacing radio as the most powerful promotional tool in music" wrote Lucas Shaw in Bloomberg Businessweek (2018, para. 1), and this perception of a powerful role of playlists in promotion and discovery was supported by interview data. Artist manager Monica's support of this narrative was typical of how participants viewed playlisting in 2017–18:

> Spotify is so huge for music discovery now, because that's how people are finding out about new music instead of listening to radio. Spotify has become this place with all the generated playlists, and they just keep feeding out playlists to the masses. That's how people are discovering their new music (Monica, personal communication, 2017).

Monica's perception of how Spotify playlisting was becoming the dominant form of music discovery was supported by other participants, regardless of whether they personally listened to them. The new perceived supremacy of playlists to promote songs was noted and echoed by the global press (Forde, 2017). This perceived transition from playlists as tools for users to organize their consumption to a powerful taste-making vehicle can be summarized by Carl Chery, then an Apple Music playlist curator who has since moved to Spotify: "I'm getting to a point where I'm so far ahead of things that I'm actually able to shift them. Instead of reacting to what's happening, I have a hand in shaping what's happening" (Ugwu, 2016).

The narrative of playlists becoming the new radio, or the most important method with which to promote songs, is indicative of the perception that playlists have become intermediaries, able to influence and affect popular taste and consumption. This metaphor of streaming being a *new radio* was mentioned by managers, labels, and publishers, and I heard it often at industry conferences. While interview data confirms that human curators handled the initial decision to add new tracks to playlists, the exact role of humans vs. automated algorithmic systems in the movement of tracks within the playlist ecosystem is opaque. Participants who pitch playlist curators are unsure and uninformed about how Spotify decides to move or drop recordings within the playlist ecosystem. A story in WIRED magazine reported that Spotify's metric evaluations of a performance of a song will largely determine what happens once it is in the ecosystem (Pierce, 2017). Both human and algorithmic forces are reportedly at play when deciding what happens to tracks in the playlist ecosystem, and this was confirmed by Sarah Hamilton, operations manager for aggregator Ditto Music.

> "The editors for the different genres at Spotify will often put a song on New Music Friday to test it and see how people react to it, and then move it on from there … I think they're seeing how people respond to music by looking at data and analytics" (Sarah Hamilton, personal communication, 2017).

Hamilton's role at Ditto music was just one example of a new position that has emerged in the wake of the growth of Spotify and other DSPs: aggregators were offering independent artists the kind of promotional services which were traditionally handled by record labels. While Ditto Music allowed anyone to use their distribution service to deliver music to multiple DSPs, the bespoke playlist pitching service was reserved for artists which Ditto selects. This was also the case with MGM Distribution.

> Because we're dealing with independent artists, some of them don't have labels. Some of them don't even have management. I personally take on the role of pitching them to the different services (Beth Deady, personal communication, 2017).

The aggregators asked for a share of revenue in exchange for this service. This shift to a service model, where marketing and promotional services are provided for a split of the revenue – as opposed to the traditional exchange for intellectual property rights given to record labels – is an interesting trend in line with the more mythical promises of the internet to create better conditions for creators.

While "gut feeling" and the perceived ability of a song to "connect" are cited as the reasons particular songs are added (Hamilton, personal communication, 2017), the cultural and social capital of stakeholders like Hamilton and Deady also play a role in their ability to get songs onto playlists. They act as gatekeepers within their own clientele, then pitch their selections to the gatekeepers at the streaming services. In interviews, it was made clear that human ears selected which songs get added, though metric evaluation of the song's performance – skip rate, shares, saves, and other responses by the user – all affected placement of the track in an opaque manner. This dynamic re-evaluation is exemplary of Powers' concept of *intermediation* as a process.

It was the cultural and social capital that professionals like Hamilton and Deady have generated through their work in the digital music sector which allowed them to hold positions to assist songs which they selected to enter the playlist ecosystem. This is also a continuation of existing recording industry logics, where the reputation of individuals as being able to hear and predict hit recordings is built up and leveraged in order to promote artist brands (Dannen, 1990; Negus, 1999). This idea that access to playlists is interme-diated by skilled experts is another similarity to the radio format that makes playlisting appealing to the recording industry: access is a skill which leveraged cultural and social capital, and this skill can be developed.

Afterwards, allegedly metrics and algorithms take over and performance replaces intermediary capital as grounds for action. Soliciting playlists was not a very different procedure than how pitching radio, press, or any other gatekeeper tends to work: personal relationships are developed over time, earning certain stakeholders better access to curators through repeatedly proving an ability to be effective at selecting recordings which perform well. Managers, labels, distributors/aggregators, and publishers informed me that they could pitch songs to Spotify's curators during this time by emailing a form, but publicists reportedly were not welcome.

Accusations of *playola* – paying to have songs added to a playlist – circulated in this period, as documented in trade magazine Billboard. However, the report did not specify clearly whether in-house curated or third-party playlists were the topic, and the playlist promotion companies described in the article were third-party companies owned by the major record labels (Peoples, 2015). Despite this ambiguity, subsequent news articles about Spotify's playlists emphasized the narrative of curation decisions being largely metric and performance-based (Pierce, 2017; Ugwu, 2016). Journalist Liz Pelly reported that major labels controlled most of the editorial content and playlist slots, while also reporting on the ability for rights owners to pay to have sponsored songs appear on playlists (Pelly, 2017). All participants denied hiring playlist promotion companies or paying for playlist placement.

Despite journalistic reports of possible manipulation and a dominance of the major label artists within the playlist ecosystem, the lack of clarity on how Spotify was actually making their decisions was troubling and confusing participants more than what the major labels or playlist promotion companies were up to. Several artists speculated that even "an algorithm is also going to favor major labels' artists, or whatever, because of certain money going into it, right?" (Dan, personal communication, 2017).

While Apple Music uses more human curation as point of difference in their brand (Morris & Powers, 2015), the more algorithmic quality of the highly-influential Spotify playlists was on everyone's mind, even if they did not understand how it worked. As Brian, the director of a publishing company with several decades of experience noted:

> The access to those playlists is becoming the new radio, or television … But now, the playlists are clearly the way the game is, and that is open to manipulation. Now, I'm not saying that Spotify is manipulating, but the way these lists are being created is not entirely clear (Brian, Publisher, interview, 2017).

While Brian was well aware that playlist pitching involved human curators, it is the opaque sociotechnical nature of how the ecosystem was subsequently managed that was concerning to him. The nuances of how the opaque algorithmic structures on Spotify functioned were not well-understood, and the platform did not share information about why decisions regarding song placement were made.

Nearly all participants were aware that gaining access to the playlists involved a pitching procedure, though the less experienced artists were unsure who they would need to engage to help them with this. Ronnie Frew from rock band Pink Harvest demonstrates the perception of a young artist without an experienced manager or label informing them as to specifics. After describing how their band receives mass mailed emails from companies promising to help them with playlist placement or boosting metric numbers, Frew notes that any kind of help requires hiring people, or spending money.

> My general opinion is that there's a money stream, no matter what. You might do it the moral way, or the way the industry considers moral. You're still probably paying something to do it. You might pay a publicist a certain amount of money that probably will get the same amount of views and they are technical views of actual people, but there's always money that starts it. It starts the process, unless you spend a lot more time (Ronnie Frew, personal communication, 2017).

Frew's words represent those of someone who knows that access will require capital and the support of other human intermediaries to gain access to the sociotechnical playlist environment, though the specifics remain unclear. While experienced playlist pitchers like Deady and Hamilton and industry veterans like Brian are aware of how to leverage their cultural and social capital to help songs gain access, the details of how the playlists are subsequently managed are opaque.

Playlist pitching became less mysterious in July, 2018, when Spotify announced that all playlist pitches would subsequently need to originate from the artist or label's account (Spotify, 2018b). Playlist pitching was thereby reframed as no longer an exclusive and mysterious domain for industry insiders. While nothing prevents pitching experts from continuing to solicit curators via the Spotify interface, the change is one of several announcements by Spotify in 2018 which appear on the surface intended to disintermediate existing intermediaries. Overall, these announcements indicated a goal to create more direct relationships between artists and Spotify.[7] However, existing relationships between curators and pitchers would not necessarily be affected by this change in pitching procedure.

The lack of clarity and information regarding how playlists were being managed was frustrating to industry professionals, in particular those managers, label staff, and publishing professionals who had more experience with having their songs on playlists. Possibly unfair access to lists through power and/or money were in comparison only a slight concern mentioned by artists. While several artists, managers, and small labels expressed tension around focusing too much on Spotify's metric and data-driven tools, not a single respondent denied the perceived power of playlists to generate streams. The emerging key role of playlists as influential sociotechnical gatekeepers was not questioned. Access to playlists was still seen as reliant on humans pitching to other humans, while the role of algorithms in managing playlists was opaque and mysterious.

Playlists and engagement: who was that artist?

The role of playlists as a focal point for discovery of new tracks evolved out of audiences embracing the format. Mary Beth Ray's study of digital music consumption in the USA pre-dates the focus on curated playlists as promotional tools, but shows that playlists emerged as crucial for everyday listening through enabling the avoidance of the labor of active curation, as well as managing the overwhelming choice offered by the vast amount of music available (Ray, 2017). The rise in streaming revenue during this period appears directly linked to the rise of playlist usage in the minds of many, and a quantitative study by the European Commission supports this: "Getting on Today's Top Hits is worth almost 20 million additional streams, which translates to [between] $116,000 and $163,000 in additional revenue from Spotify alone" (Aguiar & Waldfogel, 2018, p. 27).

While a strong perception of playlists' value for promoting a release to generate streams and revenue was consistent within the sample, concerns about the playlist as possibly detrimental to the long-term engagement of the listener with artists was on the mind of several participants on the industry side, such as Chris Maund of large independent label Mushroom Group.

> One issue with streaming is that playlists are akin to radio. To use an overused expression, it's lean back, not lean in, and it means that a lot of people are being fed the music rather than going out and discovering it. It's having a huge impact. I think albums are a great artistic outlet, and clearly streaming is affecting that. The emphasis from a financial/release/promotion point of view is moving away from albums. While I get that is the new order, it's a shame, culturally (Chris Maund, personal communication, 2017).

While the unbundling of the album in favor of focusing on individual tracks is a relatively old story in the narratives of digital disruption, it is the additional lack of active searching for the song that Maund refers to here that is the new cause for concern. Listening to a personal collection involves interaction with an archive and active selection of which albums or tracks to play. Listening to radio is, comparatively, perceived as completely passive and non-interactive. Streaming playlists are perceived to encourage similar *passive* listening within an interactive setting where the listener still has control of selection. While radio and plenty of other forms of music placement (*e.g.* synchronization, public performance of recordings) are established passive conduits for getting songs into the ears of the public, playlists are in this case framed as an encroachment of passive listening into the previously interactive realm of personal use.

There are two highly relevant differences to listening via playlist streaming to a personal collection or to the radio: 1) with all playlists, but in particular mood- and context-based ones, the aural experience of hearing the music is often not the primary focus. Rather, the listening becomes a soundtrack, or even a functional element designed to attune the listener to the activity. This is particularly notable in the numerous playlists that are framed to be listened to while working out or studying, and 2) unlike radio, there is no voice notifying the listener of what they are listening to. The listener is required to stop what they are doing and engage the Spotify interface in order to determine any information about the track.

This concern about the relative lack of artist/fan engagement on the playlist was especially on the minds of more experienced stakeholders, those at the larger labels and publishing companies. Concerns whether this lack of engagement could mean that playlist streams are not a good metric for measuring actual artist support were vocalized by Phil, another participant with decades of experience in the Australian music industries.

> A lot of streams are driven by passive playlist listening, something which can provide an income for an artist to help sustain and grow their career. However, that doesn't mean that you're in any way building an engaged fanbase or will be breaking through in a bigger way. An artist needs to dive deeper and look beyond the numbers to understand the context and where the streams are coming from to provide a better view of engagement with their fans (Phil, personal communication, 2017).

Managers, label, and publishing professionals also noted that because of this concern about streams being an unreliable metric of engagement, an artist's follower count was seen as a stronger indicator than streams. This observation about how to interpret data seemed to be a recent insight, and best practices around how to handle the different metrics and usage data seemed to be changing quickly, especially once Spotify launched their "Spotify for Artists" portal in April 2017 (Spotify, 2017).

Stories regarding artists who had high stream counts being unable to draw the expected audience were mentioned by participants from labels and live concert firms, and a few of these stories were shared with me at conferences. By 2018, journalists and industry analysts began picking up on these concerns about playlisting, as well as the reliance on streams as an indicator of genuine fandom. "Streaming analytics stripped of the context of insight can mislead. Lean-back hits are not real hits," according to industry analyst Mark Mulligan (Mulligan, 2018), and journalist Cherie Hu wrote in November 2018 that we are entering a "post-playlist reality" marked by decreased confidence in the ability of playlists as vehicles for discovering music, developing artist careers, and/or improving the overall music industry. (Hu, 2018). While the interview sample was more focused on determining how to get access to playlists, the seeds of doubt in the vitality of curated playlists as always good for artist development can be found in the remarks of the industry veterans. This ambiguity and fallibility of streaming metrics to properly measure engagement echoes the findings of Nancy Baym about the unreliability of social media metrics for measuring genuine engagement (Baym, 2013).

Conclusion

The curated in-house playlist is the mechanism whereby DSPs such as Spotify are able to combine their affordances in promotion, distribution and retail to create an instrument for generating streams for particular songs which consolidates power and creates new sociotechnical intermediaries. This creates revenue and opportunities for promotion of recordings, though the discovery of new songs through playlists is perceived to be lacking in artist/fan engagement. Journalists and trade publications have noted the popularity of streaming and subsequent growth in revenue, questioned the fairness of their revenue rates, probed the ability of money to influence curation of playlists, and were beginning to question their limitations as tools for artist development.

The interview sample shows a general enthusiasm over growing revenue from streaming from industry stakeholders, while the artists and small labels who do not see significant revenue from streaming are skeptical that it will become as significant as physical sales as a revenue stream for them. The absence of strong rhetoric claiming that streaming is inherently anti-artist and devaluing recordings was notable during this period. Lack of clarity on how Spotify was managing their playlists was more of a concern than issues of access or unfair influence, and participants who were familiar with the pitching process confirmed that human curators granted access, while the role of metrics and algorithms in managing the playlists was opaque. The power of Spotify playlists to generate streaming numbers and revenue was never questioned by participants.

We can see how the affordances of DSPs to shape listening choices as well as generate revenue combined to make playlisting the major promotional concern of this era. Concerns regarding engagement and the long-term effects of the popularity of the playlist format on artist development were on the minds of several participants. These interviews document a period in Australia where DSPs have taken the lead in revenue generation and are consolidating power in promotion and distribution. There was enthusiasm to be found for new intermediaries, but concerns and confusion about the increased reliance on opaque algorithmic automation. The lack of artist/fan engagement via the playlist was the emerging concern.

Notes

1. The *music industries* encompass the broad monetization of music, including the recording industry, live performance, and other revenue streams. This preference for the plural reflects older definitional work (Sterne, 2014; Williamson & Cloonan, 2007) and in particular Nordgård (2018) who also views the streaming services as external actors to the music industries.
2. Self-reported artist genres (acts typically cited 2–3 genres): Indie (5), Pop (3), Rock (2), Electronic (2), Funk (1), Hip Hop (1), Alternative (1), Roots (1), Soul (1).
3. *Platform* is a term commonly applied to these companies, and Spotify itself is noted to describe itself as a *music streaming subscription service* or an *audio platform* depending on context. but the usage of the term *platform* in academia carries theoretical implications that cannot be addressed here, and will be avoided. (Eriksson, Fleischer, Johansson, Snickars, & Vonderau, 2019) work through this definitional issue in Spotify's case, and propose to use *platform* as shorthand for *action nets*, in order to "grasp this quality of companies as temporary entanglements of unlike yet related actors" (p.15).
4. Vonderau (2017) reports the usage of *DSP* referring to *Demand-Side Platforms*, but that usage is focused on a particular segment of the advertising supply chain within Spotify's platform ecosystem.
5. (Prey, 2016) provides a breakdown of how Spotify's algorithmic recommendation engine the Echo Nest generates personalized recommendations, which is a different algorithmic process than the management of the curated playlists. (Eriksson et al., 2019) lays out the broad functionality of the entire Spotify system, but neither of these sources shed more light into how in-house playlists are specifically managed beyond how Spotify reports that they are: managed by human curators, and informed by metrics.
6. *Discovery* is also an important term to problematize, the subject of a special issue in this journal in 2016. See (Nowak, 2016) on the differentiated approaches to discovery as a phenomenological act. *Discovery* here is understood as the mechanism where listeners are exposed to new recordings through recommendation, and (Kjus, 2016) notes: "The degree to which these services manage to convince artists, labels and consumers of the virtues of discovery may well be decisive for the future of the streaming model." (p. 129).
7. The announcement of the new pitching process was followed in September, 2018, by the announcement that artists could upload music directly into the Spotify system without using an aggregator (Spotify, 2018a).

Disclosure statement

No potential conflict of interest was reported by the author.

Funding

This research was supported by the Australian Government through the Australian Research Council's Linkage Projects funding scheme (project LP150100156).

Cited research interviews

Brian pseudonym (2017). Director, large music publishing company
Cuming, Lou (2017). Artist manager, Mister Management

Dan pseudonym (2017). Songwriter and performer
Deady, Beth (2017). Digital Asset Manager, MGM Distribution
Frew, Ronnie (2017). Songwriter and performer, Pink Harvest
Hamilton, Sarah (2017). Operations manager, Ditto Music
Monica pseudonym (2017). Artist manager
Maund, Chris (2017). Chief Operating Officer, Mushroom Group
Phil pseudonym (2017). Managing director, large music publishing company
Wilson, Amy (2017). Songwriter and performer, Mere Women

ORCID

Benjamin A. Morgan ⓘ http://orcid.org/0000-0003-3777-351X

References

Aguiar, L., & Waldfogel, J. (2018). *Platforms, promotion, and product discovery: Evidence from Spotify playlists*. Retrieved from https://ec.europa.eu/jrc/sites/jrcsh/files/jrc112023.pdf

Alacovska, A., & Gill, R. (2019). De-westernizing creative labour studies: The informality of creative work from an ex-centric perspective. *International Journal of Cultural Studies*, 1367877918821231. doi:10.1177/1367877918821231

Albury, K. (1999). Spaceship triple J: Making the national youth network. *Media International Australia Incorporating Culture and Policy*, 91(1), 55–66. doi:10.1177/1329878X9909100108

ARIA. (2019). *ARIA 2018 music industry figures show 12.26% growth* [Press release]. Retrieved from https://www.ariacharts.com.au/news/2019/aria-2018-music-industry-figures-show-12-26-growth

Australia, P. O. (2019). *Report on the inquiry into the Australian music industry*. Canberra. Retrieved from https://www.aph.gov.au/Parliamentary_Business/Committees/House/Communications/Australianmusicindustry/Report

Baym, N. K. (2011). The Swedish model: Balancing markets and gifts in the music industry. *Popular Communication*, 9(1), 22–38. doi:10.1080/15405702.2011.536680

Baym, N. K. (2013). Data not seen: The uses and shortcomings of social media metrics. *First Monday*. 18(10). doi:10.5210/fm.v18i10.4873

Baym, N. K. (2018). *Playing to the crowd: Musicians, audiences, and the intimate work of connection*. New York: New York University Press.

Bourdieu, P. (1984). *Distinction a social critique of the judgement of taste*. London, UK: Routledge & Kegan Paul.

Bourdieu, P. (1986). The forms of capital (R. Nice, Trans.). In J. Richardson (Ed.), *Handbook of theory and research for the sociology of education* (pp. 46–58). New York, NY: Greenwood.

Byrne, D. (2015). Open the music industry's Black Box. *The New York Times*. Retrieved from http://www.nytimes.com/2015/08/02/opinion/sunday/open-the-music-industrys-black-box.html

Dannen, F. (1990). *Hit men: Power brokers and fast money inside the music business*. Sydney, Australia: Muller.

De Beukelaer, C. (2015). *Developing cultural industries: Learning from the Palimpsest of practice*. Amsterdam, The Netherlands: European Cultural Foundation.

Dredge, S. (2013). *Thom Yorke calls Spotify 'the last desperate fart of a dying corpse'*. Retrieved from http://www.theguardian.com/technology/2013/oct/07/spotify-thom-yorke-dying-corpse

Ellis-Petersen, H. (2016). Music streaming hailed as industry's saviour as labels enjoy profit surge. *The Guardian*. Retrieved from http://www.theguardian.com/technology/2016/dec/29/music-streaming-industry-saviour-labels-spotify-apple-music

Eriksson, M., Fleischer, R., Johansson, A., Snickars, P., & Vonderau, P. (2019). *Spotify teardown: Inside the Black Box of streaming music*. Cambridge, MA: Mit Press.

Fair Music: Transparency and payment flows in the music industry. (2015). Retrieved from https://www.berklee.edu/sites/default/files/Fair%20Music%20-%20Transparency%20and%20Payment%20Flows%20in%20the%20Music%20Industry.pdf

Forde, E. (2017). 'They could destroy the album': How Spotify's playlists have changed music for ever. Retrieved from https://www.theguardian.com/music/2017/aug/17/they-could-destroy-the-album-how-spotify-playlists-have-changed-music-for-ever

Gerber, R. (2017). How the music industry is putting itself out of business. *Forbes.* Retrieved from http://www.forbes.com/sites/greatspeculations/2017/05/03/how-the-music-industry-is-putting-itself-out-of-business/

Hagen, A. N. (2015). The playlist experience: Personal playlists in music streaming services. *Popular Music and Society, 38*(5), 625–645. doi:10.1080/03007766.2015.1021174

Hesmondhalgh, D. (2009). Politics, theory and method in media industries research. In J. Holt & A. Perren (Eds.), *Media industries history, theory, and method* (pp. 245–255). Chichester, UK: John Wiley & Sons.

Hesmondhalgh, D., & Meier, L. M. (2018). What the digitalisation of music tells us about capitalism, culture and the power of the information technology sector. *Information, Communication & Society, 21*(11), 1555–1570. doi:10.1080/1369118X.2017.1340498

Hu, C. (2018). *Our new "post-playlist" reality.* Retrieved from https://www.getrevue.co/profile/cheriehu42/issues/our-new-post-playlist-reality-138493

Hughes, D., Keith, S., Morrow, G., Evans, M., & Crowdy, D. (2013). What constitutes artist success in the Australian music industries? *International Journal of Music Business Research (IJMBR), 2*(2), 61–80.

IFPI. (2019). *Global music report 2019: State of the industry.* Retrieved from https://www.ifpi.org/downloads/GMR2019.pdf

Jones, M. (2003). The music industry as workplace: An approach to analysis. In A. Beck (Ed.), *Cultural work: Understanding the cultural industries* (pp. 147–156). London, UK: Routledge.

Karakayali, N., Kostem, B., & Galip, I. (2017). Recommendation systems as technologies of the self: Algorithmic control and the formation of music taste. *Theory, Culture & Society,* 0263276417722391. doi:10.1177/0263276417722391

Kjus, Y. (2016). Musical exploration via streaming services: The Norwegian experience. *Popular Communication, 14*(3), 127–136. doi:10.1080/15405702.2016.1193183

Lalonde, P.-É. (2014). *Study Concerning Fair Compensation for Music Creators in the Digital Age.* Retrieved from Nashville: http://www.songwriters.ca/studyconcerningfaircompensation2014.aspx

Marshall, L. (2013). The structural functions of Stardom in the recording industry. *Popular Music and Society, 36*(5), 578–596. doi:10.1080/03007766.2012.718509

Marshall, L. (2015). 'Let's keep music special. F—Spotify': On-demand streaming and the controversy over artist royalties. *Creative Industries Journal, 8*(2), 177–189. doi:10.1080/17510694.2015.1096618

Meier, L. (2013). *Promotional Ubiquitous musics: New identities and emerging markets in the digitalizing music industry* (PhD Monograph). The University of Western Ontario. Electronic Thesis and Dissertation Repository. Retrieved from http://ir.lib.uwo.ca/etd/1096

Meier, L. M. (2017). *Popular music as promotion: Music and branding in the digital age.* Cambridge, UK: Polity Press.

Morgan, B. (2019). Whose tool for what purpose? The struggle for cultural industry infrastructure in Liberia. In S. Labadi (Ed.), *The cultural turn in international aid: Impacts and Challenges for Heritage and the Creative Industries* (pp. forthcoming). London, UK: Routledge. Retrieved from https://www.taylorfrancis.com/books/9781351208598

Morris, J. W. (2015). Curation by code: Infomediaries and the data mining of taste. *European Journal of Cultural Studies, 18*(4–5), 446–463. doi:10.1177/1367549415577387

Morris, J. W., & Powers, D. (2015). Control, curation and musical experience in streaming music services. *Creative Industries Journal, 8*(2), 106–122. doi:10.1080/17510694.2015.1090222

Mulligan, M. (2018). *Are record labels facing an A&R crisis?* Retrieved from https://musicindustryblog.wordpress.com/2018/07/30/are-record-labels-facing-an-ar-crisis/

Negus, K. (1999). *Music genres and corporate cultures.* London, UK: Routledge.

Negus, K. (2002). The work of cultural intermediaries and the enduring distance between production and consumption. *Cultural Studies, 16*(4), 501–515. doi:10.1080/09502380210139089

Nicolaou, A. (2017, January 17). How streaming saved the music industry. *Financial Times.* Retrieved from https://www.ft.com/content/cd99b95e-d8ba-11e6-944b-e7eb37a6aa8e

Nordgård, D. (2018). *The music business and digital impacts: Innovations and disruptions in the music industries.* Cham, Switzerland: Springer.

Nowak, R. (2016). When is a discovery? The affective dimensions of discovery in music consumption. *Popular Communication, 14*(3), 137–145. doi:10.1080/15405702.2016.1193182

Pelly, L. (2017). *The secret lives of playlists.* Retrieved from https://watt.cashmusic.org/writing/thesecretlivesofplaylists

Peoples, G. (2015). How 'Playola' is infiltrating streaming services: Pay for Play is 'definitely happening'. *Billboard.* Retrieved from http://www.billboard.com/articles/business/6670475/playola-promotion-streaming-services

Pierce, D. (2017). The secret hit-making power of the Spotify playlist. *WIRED.* Retrieved from https://www.wired.com/2017/05/secret-hit-making-power-spotify-playlist/

Powers, D. (2015). Intermediaries and intermediation. In A. Bennett & S. Waksman (Eds.), *The Sage handbook of popular music* (pp. 120–134). Thousand Oaks, CA: Sage Publications Ltd.

Prey, R. (2016). Musica analytica: The datafication of listening. In R. Nowak & A. Whelan (Eds.), *Networked music cultures: Contemporary approaches, emerging issues* (pp. 31–48). London, UK: Palgrave Macmillan UK.

Ray, M. B. (2017). *Digital connectivity and music culture: Artists and accomplices.* Cham, Switzerland: Springer International Publishing.

Rogers, I. (2013). The Hobbyist majority and the mainstream Fringe. In S. Baker, A. Bennett, & J. Taylor (Eds.), *Redefining mainstream popular music* (pp. 162–173). London, UK: Routledge.

Seaver, N. (2018). Captivating algorithms: Recommender systems as traps. *Journal of Material Culture,* 1359183518820366. doi:10.1177/1359183518820366

Shaw, L. (2018). *Spotify's playlists are more powerful than radio.* Retrieved from https://www.bloomberg.com/news/articles/2018-03-23/spotify-s-playlists-are-more-powerful-than-radio

Sinnreich, A. (2016). Slicing the pie: The search for an equitable recorded music economy. In P. Wikström & R. DeFillippi (Eds.), *Business innovation and disruption in the music industry (pp. 153-174).* Cheltenham, UK: Edward Elgar Publishing.

Spotify. (2017). *Fan insights is now 'Spotify for Artists' and available to all artists* [Press release]. Retrieved from https://artists.spotify.com/blog/fan-insights-is-now-spotify-for-artists

Spotify. (2018a). *Now in Beta: Upload your music in Spotify for artists* [Press release]. Retrieved from https://artists.spotify.com/blog/now-in-beta-upload-your-music-in-spotify-for-artists

Spotify. (2018b). *Share new music for playlist consideration* [Press release]. Retrieved from https://artists.spotify.com/blog/share-new-music-for-playlist-consideration

Sterne, J. (2014). There is no music industry. *Media Industries Journal, 1*(1), 50–55. Retrieved from http://www.mediaindustriesjournal.org/index.php/mij/article/view/17/26

Strong, C., & Rogers, I. (2016). She-Riffs: Gender and the Australian experience of alternative Rock and Riot Grrrl in the 1990s. *Journal of World Popular Music, 3*(1), 38–53. doi:10.1558/jwpm.v3i1.2272

Swift, T. (2015, June 21). *To apple, love Taylor.* Retrieved from http://taylorswift.tumblr.com/post/122071902085/to-apple-love-taylor

Ugwu, R. (2016). *Inside the playlist factory.* Retrieved from https://www.buzzfeed.com/reggieugwu/the-unsung-heroes-of-the-music-streaming-boom

Vonderau, P. (2017). The Spotify effect: Digital distribution and financial growth. *Television & New Media,* 1527476417741200. doi:10.1177/1527476417741200

Webster, J., Gibbins, N., Halford, S., & Hracs, B. J. (2016). *Towards a theoretical approach for analysing music recommender systems as sociotechnical cultural intermediaries.* Paper presented at the Proceedings of the 8th ACM Conference on Web Science, Hannover, Germany.

Wikström, P. (2013). *The music industry: Music in the cloud* (2nd ed.). Cambridge, UK: Polity.

Williamson, J., & Cloonan, M. (2007). Rethinking the music industry. *Popular Music, 26*(2), 305–322. doi:10.1017/S0261143007001262

Metrics and decision-making in music streaming

Arnt Maasø ⓘ and Anja Nylund Hagen ⓘ

ABSTRACT

Music streaming enables the tracking of listening behavior in more detail than any previous music-distribution format. While it is well known that streaming services collect troves of data, little is known about how stakeholders, including managers or label executives, make metric-based decisions and how they understand the impact of algorithms. The article uses anonymized interviews with music industry professionals, exploring how they use metrics in streaming services and examining their decision-making processes. The analysis concludes that they rely on a growing volume of data when making decisions about what to promote, and how. Nevertheless, most of the stakeholders focused on fairly simple metrics, such as salient spikes that were noticeable "at a glance." When discussing these findings, we draw attention to the reinforcing feedback loops between metrics, data-based decisions and algorithms, questioning whether datafication acts to intensify trending events and diffusion of new music.

Introduction

Over the last decade, the music industries have seen a sharp growth in music-streaming services (MSSs) such as Spotify, Apple Music, Amazon, YouTube, TIDAL, and Deezer. In 2018, these services accounted for over 50% of all recorded-music revenues worldwide (IFPI, 2019). These on-demand platforms provide consumers with access (through subscription or exposure to advertising) to every song in a music catalog for a specific period of time (Wikstrøm, 2013, p. 105). The visible interface of an MSS on mobile phones or other devices generally contains "technical features (e.g., buttons, scroll bars, stars, icons) as well as regulatory features (e.g., the rule that a personal profile is required before entering the site)" (Van Dijck, 2013, p. 31). A selection of interactive features is normally also offered that enables subscribers to share, organize (in playlists, for example), search, and otherwise be creative with their music.

In addition, and most importantly for this article, the MSS utilizes an *invisible interface* (Van Dijck, 2013, p. 31) that is hidden from users and controlled by the streaming platform owner, who can manipulate it by hiding or revealing certain contents, or features, and by arranging music assets to perform in certain ways. This aspect of the MSS includes automated recommendation systems that provide listeners with music suggestions via playlists and streams through preprogrammed algorithms. For example,

---

algorithms programmed to track popular songs (trending algorithms) and match your listening history to other users (collaborative filtering), may be used to present a user with a playlist based on a particular day of the week or likely activities during the time of day ("Friday Night Party Music"), that is not identical to what another user would be recommended at the same time and place. These recommendations impact the listener's music supply, and especially the selection of songs that are foregrounded and frequently rotated. The same impact could be ascribed to online interfaces and software offered by music-streaming providers to labels, artist, brands, and publishers. Like the consumer-facing interfaces, these B2B interfaces provide data about usage, as well as rules concerning the uploading of music files and metadata to the MSS. Overall, the algorithms and these invisible interfaces impact how music is perceived and made sense of in notable and unpredictable ways.

For stakeholders dedicated to facilitating their music's path to its audiences – that is intermediaries such as managers or labels executives – the streaming algorithms present both an opportunity and a challenge. In short, music distribution through networked media such as streaming services is characterized by inherent structures that make the resultant spread more difficult for the music industry to control (Wikstrøm, 2013, p. 6). Nevertheless, the practice of streaming has become so culturally central that an understanding of the services' ways of working is critical, as they "constitute a reinforcing feedback loop that plays a crucial role in the music industry dynamics" and give "rise to (or end) fads, brands, acts, or genres" (Wikstrøm, 2013, p. 88). The new dynamics in the relationship between listeners and music that are afforded by streaming services and their algorithms are therefore crucial to the ways in which music distribution has developed over time in the streaming culture.

These dynamics are also at the heart of this article's research, which builds upon studies demonstrating the impact of MSSs on user' behavior (e.g., Hagen, 2015; Johansson, Werner, Åker, & Goldenzwaig, 2018; Maasø, 2018) to develop a perspective on how insights into data about user behavior may also trigger new patterns of decision making by different stakeholders and actors in the industry. For these stakeholders, streaming technology supplies opportunities to reach new audiences in new markets, but it also demands new skills to respond to the new competition, which now includes global partners and a host of various technologies and platforms with unique characteristics.

Fundamentally, the distribution of music online via MSSs presents an opportunity to monitor and analyze users' consumption patterns and interactions. Such monitoring activity, here coined "datafication", has not only been important to large platforms like Facebook but also to the development of Spotify and other MSSs (Prey, 2016). In this article, we will argue that datafication and metrics are becoming more central to a range of activities *outside* of the streaming services and are of growing importance in the strategic planning and execution of efficient music distribution among various other stakeholders in the music industry. The overarching research question for the article is as follows:

> How are metrics of streaming usage influencing strategic choices and actions taken by stakeholders in the music business, and what are the relations between the metrics of streaming and the algorithmic affordances of music distribution in music-streaming services?

In terms of this investigation, the Norwegian music business is a fruitful starting point, as Norway was one of the first markets to adopt streaming, which surpassed revenues from physical sales as far back as 2011 (IFPI, 2014). It currently accounts for 89% of the total revenues from recorded music in Norway (IFPI, 2019). Spotify has always been the dominant MSS in the Nordics, and it is the service brought up by most of the informants in this study; hence, it will be the prime example in our discussion below.

In the following section, we will discuss existing research on this study's core concepts and connect them to the workings and features of on-demand MSSs.

Central concepts: datafication, algorithms, affordances

The term datafication was first coined by Mayer-Schönberger and Cukier (2013) to describe a broad range of phenomena through which our everyday life is tracked, quantified, and analyzed to inform predictive conclusions. Van Dijck (2014) discusses datafication in relation to surveillance and "life mining" or "dataveillance" by social media platforms. Couldry and Hepp discuss the impact of datafication in digital media and platforms, where "[datafication and digital traces] themselves become part of constructing media events" (2017, p. 116).

As mentioned, the MSS introduced the possibility of tracking in great detail how music is listened to in daily life by huge audiences. As part of the business model of the MMS, each and every stream (of more than 30 seconds) is counted, and these numbers directly impact how revenue is both generated and shared among the music's rightsholders. In addition, usage data, such as how many users are streaming what and where, the origination of the streams (lists, searches, browsing, and so on), skips, repeats, and so on, represents many potentially important clues to better understanding about the fate of their work for stakeholders including composers, artists, managers, and record labels.

The front-facing user interfaces of MSSs highlight selected lists or songs on the listening device. Decisions regarding what to feature here are made by both editorial teams and automated algorithms, programmed to calculate and predict data regarding anticipated usage. Understanding the algorithms and affordances of the various listening devices and platforms involved with MSS is hence crucial to understanding how MSSs work.

Some of the oft-used definitions of algorithms describe them as formal processes or sets of step-by-step procedures that are often expressed mathematically (Striphas, 2015) or encoded to transform input data into a desired output, based on specified calculations (Gillespie, 2016). However, no final agreed-upon definition of algorithms exists (Bucher, 2018), and Seaver (2017) even argues that arriving at a common definition is not important; instead, we should "approach algorithms as 'multiples' – unstable objects that are enacted through the varied practices that people use to engage with them" (2017, p. 1). Algorithms may thus be viewed as "heterogeneous and diffuse sociotechnical systems, rather than rigidly constrained and procedural formulas" (ibid.).

Seavers's insight highlights how algorithms, over the last few decades, have moved beyond mathematics, informatics, and programming to increasingly engage social, political, cultural, and epistemological concerns, which is also why they are relevant here. It is important, that is, to understand the role of algorithms in online *power relations* (Beer, 2009) and the ways in which media and technology platforms steer visibility/invisibility and attract attention (Bucher, 2012a). Similar to the way in which Facebook's algorithms boast both the power and the threat of invisibility (Bucher, 2012b), software and algorithms in MSSs have the capacity to generate attention from audiences. Hogan (2015)

claims that the dominant ideology of information management "is one of sorting, especially personalized or relevance-based sorting, and infused with faith in machine learning"
(p. 103) and shows how trending and ranking algorithms are dominating contemporary
media platforms. Similarly, Gillespie has explored how trending and ranking algorithms
make some information visible while hiding other information (Gillespie, 2016, 2018).

Understanding how algorithms work involves unpacking the affordances of the MSSs
and user interfaces through which people stream music. The term affordance was first
coined by Gibson (1986) as part of his ecological approach to human perception. Central
to the concept was a relational approach to observers and environments: "The affordances
of the environment are what it *offers* the animal, what it *provides* or *furnishes*, either for
good or ill" (1986, pp. 2432–2433). In Gibson's initial description of the approach, he
noted that surfaces afford various interactions to various observers – whether the surface
is "walk-upon-able" or "sink-into-able" depends upon both the surface and the observer in
question, including the characteristics of the environment, the capacity and need of the
individual, the context of the situation, and so forth. This approach has proven relevant to
human encounters with digital technology as well, including, in the present case, what
streaming algorithms afford in terms of users' music consumption, and what the resulting
widgets and data points afford to industry stakeholders as they make their plans.
Affordances involve multiple layers, including contexts, objects, human communication,
social relations, and non-human agency (Bucher, 2018).

Algorithmic affordances, as used in the following analysis, encompass the ways in
which music is made available to users in music-streaming services, the ways in which
this process is influenced by algorithms, and the ways in which the algorithms trigger new
patterns of music consumption and data-based decisions and action (such as whether to
invest part of a marketing budget on a given artist). Below, we will begin by describing the
informants we interviewed in the hopes of addressing these issues.

Methods

The present analysis relies on material from fifteen semi-structured interviews (lasting
from fifty-five to ninety minutes each) with anonymized music industry professionals,
including twelve men and three women. The informants were recruited between spring
2018 and winter 2019, based on their position and expertise in the Norwegian music
industry. Seven informants were music managers, most of whom also offered label,
booking, publishing, or promotion services to help Norwegian artists achieve global
penetration. The relevant management firms included between one and ten employees
and boasted rosters of one to twenty artists/bands in both mainstream and niche popular
music genres. The other eight informants were stakeholders from other parts of the music-
streaming ecosystem, all with different insights into data and metrics: three informants
with experience as employees at two different MSSs; one data analyst working at one of
the three major labels; two label executives (one from a major label and one from a smaller
indie label) who worked with or managed A&R, promotion, campaigns, and making
various strategic decision for their labels and artists; and three informants who worked
for music distributors or aggregators (such as The Orchard and Ingrooves Music), helping
artists, labels, and managers get their files onto MSSs, as well as providing insights and
analysis of streaming patterns and tracking royalties. Five of the informants had recent

experience at multiple companies and in various roles (for example, both for an MSS and for a label), and they were therefore able to provide relevant insight into several of the roles mentioned.

The informants have mainly worked with data and distribution practices in commercial popular-music contexts, but the material presented below describes how negotiations with streaming technology fundamentally inform industry developments in ways that likely also resonate with other music genres and partners.

The study's method centers on expert interviews in the interests of both exploration and orientation in a relatively new field (Flick, 2018, p. 236). The choice of a variety of stakeholders as expert interviewees was strategic, as the general need for these figures has increased amid the relative financial chaos of the streaming culture (Gordon, 2014), including new actors, tech companies, and digital platform providers (and their international partners) who pursue negotiations and regulations at unprecedentedly professional levels. The inclusion of various stakeholders and intermediaries also provided us with more insight into the various types and uses of data in the music industry, as well as a deeper understanding of how music metrics may have changed over time.

The interviews seek to capture the informants' *process knowledge*, which is instructive for understanding sequences of action and interaction in the field, as well as their *interpretive knowledge*, which addresses the expert's subjective orientations, rules, points of view, and interpretations (Flick, 2018, p. 237) as they are informed by streaming and algorithms. The analysis relies on our interpretations of the informant's accounts of their professional practices.

As a methodological framework, this approach provided us with the ability to explore the stakeholders' experiences of data-driven decisions in their work. The study hence contributes to a better understanding of the sociotechnical consequences of datafication and algorithms, as a core insight important to highlight both in the present and the future music industries. We will structure and present our results in the form of four main findings. The discussion that follows proposes a way of understanding the flow of data and decisions, and the ways in which algorithmic affordances and metrics may influence trends. Ultimately, we believe that the ways in which data are used may provide compelling insights into role of music-streaming services in music culture.

Findings

Increasingly detailed data

A first (and anticipated) finding, confirmed by all informants, is that data are indeed central to their decision-making, and that the skillset in using metrics is becoming more advanced.

> Twenty year ago, there was very little business-intelligence competence at all. I was shocked, as an economist, to enter a place where most were self-taught. The CFO I reported to did not even have a high-school diploma. [...] Today we have dedicated analysts belonging to a global analytics team with hundreds of people. (Informant 13, major label executive)

> [When] we started this small label nineteen years ago, we did not know much about what happened with the use of music at all. Now we are sitting in front of screens most of the time,

with ten to fifteen different tabs open with different streaming services and other online services that we constantly monitor. (Informant 11, small label executive)

While streaming is almost a decade old in Norway, most informants noted that it has only been since 2015 that music metrics have become sufficiently detailed for the use of artists, managers, and smaller labels. Several of them (Informants 1, 2, 4, 7, 11, 13, and 15) pointed to *Spotify for artists* (software intended for artists and managers launched in April 2017 following the release of beta version called *Fan insight* in November 2015) as comprising a milestone in music metrics (Spotify, 2017), as well as *Spotify analytics* (software intended for labels and distributors). A former executive at a competing MSS described a lack of business-intelligence resources when they initially built and launched the service: "The first few years, all that mattered was to model the revenue shares, and to report this and the market share to record labels and CMOs [collective management organizations]" (Informant 10, MSS executive).

Today, music metrics are essential to music managers and their respective artists and collaborating teams. Retrieving and analyzing data from streaming services and other platforms have become critical ongoing assignments that strongly influence everyday decision making.

> My job is about getting up in the morning, and then I log on to see the analytics [...]. It's like a speedometer that we didn't have before the streaming economy, and it's fantastic. The work is simply to follow the numbers and try to understand what's going on. Because, if something starts to simmer, then we ought to see it, turn around, get geared up to act and just do it – maximize it. That's the key, really, even in studio work and A&R-ing. It's all about getting a grasp of the market's responses to the music. (Informant 4, manager)

This quote specifies that some kinds of interpretations and actions are made possible through insights from data. Datafication is decisive for strategic and relational planning, communication, cooperation, and the execution of music launches, marketing, publishing, audience targeting, and approaches to new and emerging markets. According to two of the informants (11 and 13), some labels are even using machine learning in their decision making regarding which artists to sign.

Metrics influence some of the managers' interpretations of data related to the long-term development of artist, but also their attempts to predict success or mitigate risk in the short term. For the latter, for example, artistic or commercial content can be pushed online to trigger the audience's interest before a concert or new release, based on existing metrics, and then this gesture will generate new metrics as well:

> We don't even have to release a song to test the audience interest. Instead, we can see whether we get likes on a picture. As such, streaming numbers, ticket sales, and followers on Instagram are all parts of the same picture. (Informant 1, manager)

The manager's observation illustrates datafication's permeation across the music business and from different services. In the interview, he goes on to argue for the economic benefits of metrics for his small company, given that large investments and binding agreements with other parties are less relevant today than they were for pre-digital processes of record production and distribution. His small management firm is now able to monitor market responses, streams, sales, and general interest using data from various sources, offering

him more control over his business because the risk is reduced through content development (e.g., music releases) and strategic planning (e.g., tour planning).

In another example, interpretations of data made a difference in the development of the
careers of two (here anonymized) artists.

> Take the case of artist X versus artist Y. X is an EDM artist, while Y is a more distinctive,
> quirky pop artist. X has more monthly listeners that Y, but Y has close to 500,000 followers in
> the streaming service. X doesn't even have 10,000 followers. Nevertheless, X has streams,
> because X's music spreads more naturally to playlists with many followers. (Informant 6,
> manager)

This manager uses streaming numbers to inform career moves for the artists they
represent, "and Spotify is particularly good at making this [information] available"
(Informant 6). The manager explains that the teams working with the artists, which
include label, booking, production, and promotion partners, also are more likely to
agree to strategic decisions when they are supported by data. In the case of artist X, the
data would most likely support both a production esthetics and a market drive toward the
goal of sustaining X's success on the algorithmic playlists he has already performed well
on. The larger goal is to reach thousands of followers and earn streaming revenue. For
artist Y, on the other hand, the high *following numbers* for her streaming profile suggest
that more people are aware of her music via artist recognition than via the random
streams generated through algorithms and playlists. Given this information about her
legions of followers, the managers correctly predicted and prepared for a significant
commitment to live performances for artist Y. Supported by information about the
listeners' age, gender, location, and so on, the managers accurately identified audience
loyalty and undertook successful tour planning and profitable concert deals all over the
world. The fact that this informant's strategic thinking was so heavily informed by data
suggests that industry partners are willing to adapt their work to the algorithmic system of
the MSSs. Put differently, the rationale behind these career management moves is a result
of both datafication and the algorithmic affordances that inform investments, activities,
and even production esthetics.

In general, the informants had plenty to say regarding digitization's influence on their
professional practice, particularly in terms of dealing with data. They also appeared open-
minded and willing to learn. Interestingly, the things they were *less* capable of seeing and
influencing also came up. Shortcomings were often expressed in relation to the workings
of algorithms. Several managers, for example, identified challenges around controlling
music-streaming distribution simply because the algorithms were so diffuse, functioning
as they did in opaque, complex systems. Similar views were expressed by two of the
informants who had worked in editorial teams at two different global MSSs (Informants
12 and 14). Both expressed frustration with their lack of understanding of the technical
side of algorithms, and challenges associated with cooperating with the programmers
making the software used to monitoring streaming: "I don't even know if the coders are
sitting in the US or India," said Informant 12.

The overall attitude amounted to a clear acknowledgment of the importance of algorithms for song performance, with managers further mentioning algorithms' influence
upon streaming numbers and wondering how to explain dips in popularity and changes in
playlist positions and so on. Their interpretations of their own interactions with

algorithms in MSSs were couched in terms of failure or success through a kind of experience-based recognition that they would build upon in their future practices related to music distribution.

The competitive advantage of data

A particular finding of interest to this special issue, is that all the informants highlighted Spotify's competitive edge, not only concerning marked share, but in providing useful metrics.

One manager illustrated this while showing us *Spotify for artist* on his mobile:

> "[W]e have all the information here at the tip of our fingers. Google also provides good data, but they don't have a dedicated app. […] TIDAL used to be the best the first years, but has been a black hole after they changed owners [in 2015] […] It is easy to prioritize the ones that are the easiest to work with. And Spotify is clearly the most accessible" (Informant 2, manager).

Clearly, Spotify has been developing good relations with stakeholders over time, increasing its use value through, among other things, its advanced metrics:

> Spotify is increasingly providing strategic data – relevant data points and analyses that make it possible to take action. (Informant 8, music distributor)

Spotify's acquisition of the leading analytics company Echo Nest in 2014 (Prey, 2016) was also mentioned by several informants (8, 9, and 13) as important to improving the quality of its music metrics.

Still, despite all of the data Spotify collects in house, it still turns to other stakeholders for access to *other* data points that it does not have. When managers, labels reps, or distributors are pitching new artists or music for playlists, Spotify face what are known as "cold start problems" (Seaver, 2018), because it simply lacks the data to inform its algorithms' predictions and choices. As one informant explains, "What Spotify, Apple, and others say [in pitch meetings] is 'Why should we push this particular artist? Do you have any metrics indicating that it would be useful for us to do this?'" (Informant 11).

Stakeholders then produce a host of data sources including radio play, social media presence, and tour success for Spotify, amounting to an exchange of data, often in face-to-face meetings, that builds relationships among the parties over time. Other moments of exchange and relationship building include music business gatherings or face-to-face sessions with stakeholders:

> We hosted many meetings between small and large stakeholders and Spotify, where they built trust through answering questions and being transparent (with data). (Informant 9)

When Spotify also opened several APIs for third party developers and stakeholders like music distributors and labels in 2013–2014, and more data about artists in 2013 access to data over time contributed to strengthening the relationship and reliance on Spotify, over their competitors – as was clearly their intention (Dredge, 2013).

Events

Most MSSs feature different types of *events* on the user interfaces of their consumer-facing software. These events might be created by the services' editorial teams – that is, employees with the power to curate and control service content – or fueled by algorithms. An informant with a background on editorial teams at two competing MSSs explained:

> We would typically have weekly meetings with colleagues in the other Nordic territories and discuss what worked in international and local markets, based on the data we had. We would discuss trends, brainstorm campaigns, [think about] which events and seasonal happenings we could tag along with – graduation parties, "after ski" season, summer and festivals, and so on – and we would then create playlists and campaigns based on these events. (informant 12, editorial team at MSS)

The biggest events typically claim priority real estate on the service's interface – for example, in the top banner "announcement" on Spotify's desktop version or the top row of the "home" tab. Often they resonate with general public events that are already absorbing people's attention, such as the holiday season or a major music festival. Other events start out as unplanned happenings, such as commemorations of the deaths of popular icons like David Bowie or Prince (Maasø, 2018). Memorial playlists and events are also prioritized on the MSS' interfaces and generate notable streaming numbers.

Important recurring events in the music industry, such as album releases, also receive attention from the MSSs. In 2015, a globally synchronized adaptation of the recording industry to the digital music industry produced an agreement to move the new music release day from Tuesday to Friday (Grow, 2015) – perhaps to better accommodate streaming patterns, and present fresh music for the Weekend? (Fridays and Saturdays have dominated in usage since the launch of MSSs, cf. Maasø, 2014). Playlist events such as Spotify's *Release Radar* and *New Music Friday* promptly followed, gathering a selection of new releases into playlists every week. Industry-related playlists and events are often followed with great interest by general audiences, artists, A&R reps, promoters, and booking agents, as well as the intermediaries of this study.

Moreover, playlist updates have become events in and of themselves, cementing the crucial role of the MSS in contemporary music distribution. The attention these playlists generate reinforces both their positions in the MSS and the MSSs' position as gatekeepers worth following, because they curate "the best released tracks collected in one playlist," as Spotify claims.

Because MSSs are considered the gatekeepers of new music, stakeholders, including our informants, work hard to nourish their connection. One said: "I try to keep in close contact with Spotify, several times a week, if we have forthcoming releases. It's basically: 'Can you add it on New Music Friday?'" (Informant 4, manager). Moreover, the MSS affordance of event-driven music distribution motivates pitching, marketing, and release coordination with event playlists in mind, simply because an invitation to these prioritized events can kickstart streaming of one's music.

New Music Friday and other lists are therefore regarded almost like "release parties," or "a place we can perform very well and get to the top" (Informant 1, manager). The opposite is also true: "If some person at Spotify decides not to [playlist you], the track falls completely out of the loop" (Informant 4, manager). Missing out on a playlist hence has consequences for the algorithmic loops that define the distribution logic and trending algorithms (Gillespie, 2018).

Event-driven music distribution seems to have become more important in recent years, presumably because machine learning has improved and the user base of the MSSs has increased dramatically, thereby increasing the number of data points feeding the algo-rithms. Major streaming events thus seem to spread even more virally around the world, serving to propel a few blockbuster tracks to the tops of ranking and trending lists, searches, and playlists. Spotify and other MSSs also provide playlists of top-ranking songs nationally, globally, and virally, completing the feedback loop between what is promoted and what is listened to and what, in turn, drives further promotion and listening through algorithmic success. Monitoring such data points can also fuel action, as one manager explains:

> Once we have reached a certain level, it's all about using the numbers to reach even further. To the Global Top 50 [playlist], for example. We might see that we lack 50,000 streams. Then we try to influence this by approaching a playlist in a smaller market – obtain streams over there, then hit the bigger playlist. (Informant 5, manager)

A similar approach is taken when using metrics in search engine optimization and to manipulate search patterns:

> Often, we make marketing decisions to direct users to a playlist rather than to a track, because if the data shows that the playlist will get X number of additional streams, then the list will go to no. 1 in search within Spotify. Which will then get you organic traffic. (Informant 15, data analyst)

Fresh data and novelty

Another interesting pattern involves the informants' regular reliance on real-time data – that is, up-to-the-moment metrics focused on "spikes," or sudden or salient changes. Even managers invested in understanding the long-term consequences of streaming for the artists they represent gauge those consequences via the "speedometer" of real-time changes. One seasoned executive said, "If [the data] is not provided in real time, it is rarely used. People don't go back in time to interpret data. They are interested in the present." (Informant 14, music distributor). Several other informants echoed this senti-ment, using formulations such as "here, I access the day-to-day data" (Informant 7), "the focus is on the here and now" (Informant 9), and "snapshots in real time at a glance – that's what we want from data" (Informant 11). Real-time metrics provide the opportu-nity to act quickly. For instance, informant 2, who was quoted earlier discussing the competitive advantage of *Spotify for artists*, went on to explain why fresh data were so important:

> It is difficult to prioritize cooperation [with an MSS] when you do not know the effect of your activities. […] If we've made a marketing effort, have a good story, have done something extra – [with MSS TIDAL at this time] we still have no idea about the effect. Today, we perhaps only get the data after seven months. It's hard to tell whether an activity has had any effect at that point. (Informant 2, manager)

Fresh data are hence seen to provide "instant feedback" to stakeholders regarding whether their activities and choices are working.

A couple of informants also mentioned software that provided *automated* detection of spikes including Informant 12 (who worked for an MSS) and Informant 14 (who worked for a music distributor): "We use 'spike notifications' actively, which trigger a message that is sent directly to the labels."

When prompted, a few informants acknowledged that they occasionally review old data. Informant 11 brought old data to meetings with Spotify when he was pitching an artist who had a new release after a long hiatus, and he searched old data for a "hook" to get the local press to cover a new release.

Among the music managers, only one (Informant 4) talked about occasionally digging deeper into old data, using *Spotify analytics* and the subscription service *Spotontrack* to "follow the journey" of a particular song since its release, including how many playlists it had been added to and when. Yet these exceptions to the fresh data rule did not fuel important data-based activities or decisions, except in the case of the one data analyst among our fifteen informants. While he also usually dealt in fresh data ("four-year-old data is useless"), he acknowledged that he explored historical data in order to understand how the algorithms of a particular platform work (such as the search and recommendation algorithms on YouTube) and to investigate allegations of fraud and so-called "fake artists". He also drew from both old and new data while exploring the emergence of K-pop as a genre:

> What I did was to try to "reverse engineer" how [anonymized band] spread through the internet, in order to see if we could replicate this. [...] So now I'm talking every week with the Korea office, because I did this analysis. (Informant 15, data analyst at a major label)

Exploring and acting upon old data patterns that are not easily detected by the readymade metrics highlighted by the MSS dashboards require particular resources, time, and skills in coding, and sometimes the collaboration of a global team of experts. Only the largest labels and distributors, as well as the MSSs themselves, have these kinds of resources, as our informants acknowledged:

> Our challenge is that we cannot hire statisticians and data analysts, which is what Universal, Sony, and Warner do. They have the resources to hire people who crunch data. As a small label, we have to figure this out ourselves. The major challenge is not the amount of data we get, but getting something useful. It is so much data! So many different things! (Informant 11, small-label executive)

Most stakeholders, that is, are forced to focus on the present, the trending, and the relatively obvious, whether they want to or not.

Even efforts related to making *old* tracks popular to new audiences are generally informed by real-time music metrics:

> Almost by accident, we had a track in SKAM [the TV series], which made us turn around quickly to maximize the track. The song, which already was a few years old, suddenly became super relevant again. And it also opened up new territories and markets. In Sweden, it was up on Top Three in Spotify. [...] So, this "renewal" thing, it changes the work of marketing music as a product, because nothing is ever dead. It can always be made relevant again. (Informant 4, manager)

This illustrates the impact of real-time metrics on understanding the immediate potential of monetizing a track. Informant 4 went on to explain that these numbers even made the argument for setting up a concert tour in Sweden with the artist whose track had bounced back.

The focus on novelty in the music industry is by no means unique to a streaming context, but its relevance to current music distribution has been enhanced by the vast networks of listeners who are affected by the MSS algorithms and those stakeholders who both monitor and feed those algorithms via data-based action, as we will address below.

Discussion: reinforcing feedback loops and algorithmic affordances

As we have seen, over the last few years, streaming services, and Spotify in particular, have made an increasing amount of data available to different stakeholders on a daily basis, and those stakeholders have enjoyed increasing flexibility to monitor and act based upon these metrics. No single stakeholder has access to all relevant data points about music, but it is clear that the MSSs and major labels are dominating the metrics race. These actors also have the resources and skillsets to interpret this data in ways that others do not.

According to our informants, Spotify has been the most successful actor at acquiring and providing music metrics and has capitalized on its first-mover advantage in the music-streaming business even in the company of such data behemoths as Amazon, Google, YouTube, and Apple.

Based on the way the informants use metrics, we realize that datafication in the music business is *relational*, flowing between different stakeholders and services in multiple directions. The true power of MSSs resides in the way in which they have become central information hubs, with links and feedback loops to all of the other stakeholders in the music business – either directly, through the interfaces and algorithms they control, or indirectly, through data gathered from partners and intermediaries.

We also propose that datafication functions *cyclical* or *processual*: action taken by a stakeholder or MSS creates data and metrics to be again interpreted and reacted to – creating reinforcing feedback loops of action and reaction. Similar to how Couldry and Hepp (2017) discusses datafication and media events, datafication itself may create trends and influence the music culture thanks to the algorithmic affordances of MSS. Since there are millions of actors in this information network, many people may interpret the salient signals in the same way, instigating similar actions and reinforcing trends and network effects. Data, then, beget data.

Of particular interest in terms of the broader implications of music metrics are the choices made by algorithms regarding what to render visible or invisible. Algorithms are clearly programmed to show or hide music in the interfaces used by regular listeners to stream music, as is the case with algorithm-fueled content on other social media platforms (Beer, 2009; Bucher, 2012a). Music streaming services also rely on algorithms to power the software facing artists, managers, labels, distributors, and other insider stakeholders, supplying them with metrics on trends, top lists, follower numbers, and so forth. As Hogan (2015) has shown, *sorting* and *ranking* ultimately dominates the way in which software engineers design and program interfaces. Based on the interviews, and observing the different software in use, this is the case for MSSs as well. The ideology of sorting is echoed in important ways by our interviewees' framing of their situation – how they talk about, understand, and act upon data, and why they focus on novelty, lists, trends, and spikes.

When metrics are presented, interpreted, and acted upon by numerous stakeholders making decisions that is then fed back to, and doubly reinforced by, algorithms in both consumer and business facing software, we suspect that these reinforcing feedback loops may have cumulative effects over time that in turn may impact the diversity of music culture writ large. Datafication, then, risks reinforcing only the most salient data, thereby becoming a tool best suited to making superstars and global mega-events even bigger, at the expense of smaller acts and local events.

Acknowledgments

This research was supported by The Research Council of Norway, grant 271962. The authors thank two anonymous reviewers for valuable comments of earlier versions of this paper. We also thank copy editor Nils Nadeau for substantially improving the language, and colleagues in research projects *Music on demand: Economy and copyright in a digitised cultural sector* and *Streaming the culture industries* for useful support and comments.

Disclosure statement

No potential conflict of interest was reported by the authors.

Funding

This work was supported by the Research Council of Norway [271962, 263076].

ORCID

Arnt Maasø   http://orcid.org/0000-0003-1046-4178
Anja Nylund Hagen   http://orcid.org/0000-0002-5802-2865

References

Beer, D. (2009). Power through the algorithm? Participatory web cultures and the technological unconscious. *New Media and Society*, *11*(6), 985–1002. doi:10.1177/1461444809336551

Bucher, T. (2012a). Want to be on the top? Algorithmic power and the threat of invisibility on Facebook. *New Media and Society*, *14*(7), 1164–1180. doi:10.1177/1461444812440159

Bucher, T. (2012b). A technicity of attention: How software "makes sense". *Culturemachine. Net, 13.* Retrieved from http://www.culturemachine.net/index.php/cm/article/download/470/489

Bucher, T. (2018). *If … then: Algorithmic power and politics* (Oxford Studies in Digital Politics). Oxford, UK: Oxford University Press. Kindle Edition.

Couldry, N., & Hepp, A. (2017). *The mediated construction of reality.* Cambridge, UK: Polity Press.

Dredge, S. (2013, December 3) *Spotify opens up analytics in effort to prove its worth to doubting musicians.* Retrieved from https://www.theguardian.com/technology/2013/dec/03/spotify-analytics-musicians-streaming-music-artists-earn

Flick, U. (2018). *An introduction to qualitative research* (6th ed.). London, UK: SAGE.

Gibson, J. J. (1986). *The ecological approach to visual perception.* London, UK: Lawrence Erlbaum.

Gillespie, T. (2016, June 23). The relevance of algorithms. *Culture Digitally.* Retrieved from http://culturedigitally.org/2012/11/the-relevance-of-algorithms/

Gillespie, T. (2018). #Trendingistrending: When algorithms become culture; Social media collective. In R. Seyfert & J. Roberge (Eds.), *Algorithmic cultures: Essays on meaning, performance and new technologies* (pp. 52–75). London, UK: Routledge.

Gordon, S. (2014). What is a music manager? Here's everything you need to know. *Digital Music News.* Retrieved from https://www.digitalmusicnews.com/2014/08/28/now-know-everything-music-managers/

Grow, K. (2015, 26.2). Music Industry Sets Friday as New Global Release Day. *Rolling Stone.* Retrieved from https://www.rollingstone.com/music/music-news/music-industry-sets-friday-as-new-global-release-day-73480/

Hagen, A. N. (2015). *Using music streaming services: Practices, experiences and the lifeworld of musicking* (PhD dissertation). Faculty of Humanities, University of Oslo, Norway.

Hogan, B. (2015). From invisible algorithms to interactive affordances: Data after the ideology of machine learning. In E. Bertino & S. Matei (Eds.), *Roles, trust, and reputation in social media knowledge markets* (pp. 103–117). New York, NY: Springer International Publishing. doi:10.1007/978-3-319-05467-4_7

IFPI. (2014). *Det norske musikkmarkedet 2014. Årsrapport.* Retrieved from http://www.ifpi.no/flereVnyheter/item/79VdetVnorskeVmusikkmarkedetV2014

IFPI. (2019). *Musikkåret 2018.* IFPI. Retrieved from http://www.ifpi.no/musikkaret-2018

Johansson, S., Werner, A., Åker, P., & Goldenzwaig, G. (2018). *Streaming music: Practice, media, cultures.* New York, NY: Routledge.

Maasø, A. (2014, February 27). *Music streaming in Norway: Four trends four years later.* Presentation at By: Larm conference. Retrieved from hf.uio.no/imv/forskning/prosjekter/sky-ogscene/publikasjoner/bylarm14-cloudsandconcerts.pdf

Maasø, A. (2018). Music streaming, festivals, and the eventization of music. *Popular Music and Society, 41*(2), 154–175. doi:10.1080/03007766.2016.1231001

Mayer-Schönberger, V., & Cukier, K. (2013). *Big Data: A revolution that will transform how we live, work, and think* (Kindle ed.). Boston, MA: Houghton Mifflin Harcourt.

Prey, R. (2016). Musica analytica: The datafication of listening. In R. Nowak & A. Whelan (Eds.), *Networked music cultures* (pp. 31–48). London, UK: Palgrave Macmillan UK.

Seaver, N. (2017). Algorithms as culture: Some tactics for the ethnography of algorithmic systems. *Big Data & Society, 4*(2), 205395171773810. doi:10.1177/2053951717738104

Seaver, N. (2018). *Captivating algorithms: Recommender systems as traps (prepublication).* Retrieved from https://static1.squarespace.com/static/55eb004ee4b0518639d59d9b/t/5b707506352f5356c8d6e7d2/1534096646595/seaver-captivating-algorithms.pdf

Spotify. (2017, April 19). Fan Insights is Now 'Spotify for Artists' and Available to All Artists. Retrieved from https://artists.spotify.com/blog/fan-insights-is-now-spotify-for-artists

Striphas, T. (2015). Algorithmic culture. *European Journal of Cultural Studies, 18*(4–5), 395–412. doi:10.1177/1367549415577392

Van Dijck, J. (2013). *The culture of connectivity: A critical history of social media.* Oxford, UK: Oxford University Press.

Van Dijck, J. (2014). Datafication, dataism and dataveillance: Big Data between scientific paradigm and ideology. *Surveillance & Society, 12*(2), 197–208. doi:10.24908/ss.v12i2.4776

Wikstrøm, P. (2013). *The music industry: Music in the cloud* (2nd ed.). Cambridge, UK: Polity Press.

Digital music gatekeeping: a study on the impact of Spotify playlists and YouTube channels on the Brazilian music industry

Dani Gurgel, Luli Radfahrer, Alexandre Regattieri Bessa, Daniel Torres Guinezi and Daniel Cukier

ABSTRACT

This research analyzes the audience growth of some Brazilian musicians belonging to different music genres in a 12-month period and attempts to identify the path taken through the music consumption platforms, such as YouTube channels and Spotify playlists. It also discusses the influence of strategic professionals and advertising pushes towards gaining and sustaining popularity. We rely on contemporary gatekeeping, *datacracy*, and platformization theories along with current music business and digitization studies, in order to include new stakeholders of a digitally distributed music landscape and present a proposal of *digital music gatekeeping* process. The analyzed musicians belong to very specific genres and cultural contexts. Nevertheless, it is suggested that some of the popularity strategy patterns identified by this research can be easily adapted to other cultural contexts and music genres.

Introduction

The music industry, like most of the entertainment industry, is shaped by its audience, following commercial interests and personal preferences. While the former seem quite straightforward (record labels need to be profitable in order to fund their operations), the latter are quite unclear. Since it seems quite reckless to build businesses upon transient whims, the music industry appears to have been trying to influence mass media audiences towards artists of commercial value.

This "taste influence" makes the entertainment business singular, for it has to blend together and synchronize some interests that may be quite different, even opposite: business executives with tight deadlines demanding commercial results have to negotiate both with artists that may have subjective approaches and creative choices, and with media companies that may follow popular demands (together with their own commercial interests) in order to reach more prosperous compromises.

After a lot of trial and error, these parties seemed to have got to some sort of a delicate balance by the end of the 20th century. A balance which they tried to sustain at all costs, by even attempting to charge royalties from cassette tape manufacturers to remedy the lost sales of copied albums (Drew, 2014). This landscape changed with the popularization of the Internet and social media, leading to a much greater offer of listening devices and services, more music available for the public, and more artists with access to recording studios, both professional and home assembled (Leyshon, 2014), and digital music distribution services.

We propose to study how music reaches its listeners through this new streaming environment, in order to trace a path from its production to the listening of it. Until circa 20 years ago, mass media was expansive, originating from a central point (McLuhan, 2001), and our premise is that also was the case with major label produced music. Our communication processes were constrained in what Sauerberg (2009) and Pettit (2012) have named *the Gutenberg parenthesis* – an exceptional period marked by the invention of the printing press by Johannes Gutenberg up to the legitimization of sampling, remixing, reframing culture, which then led to what they name *secondary orality*. When applied only to music, it brackets an exception to humanity's natural orality and live music in which our perception of the validness of content derived from the wideness of its circulation. Because they are seen as *official* in the parenthetical world, larger media outlets got more credibility, just as larger recording companies, also known as *the majors*. The closing of the *parenthesis* made the *secondary orality* possible, in which the sharing of music among its listeners is no longer a specificity of rare niches or a special moment such as producing a *mixtape*, but perhaps the main device in the growth of an artist's online presence.

The popularity of the Internet, broadband communications, mobile devices, and social media led to a major transformation of the music business environment. It is not an understatement to say that audiences are better connected and more informed, that there are multiple new music outlets, and that artists can now have direct communication channels with their audiences (De Marchi, 2018). Fans are no longer geographically limited as they gather in online communities and exchange musical recommendations (Baym & Ledbetter, 2012), while also pursuing direct communication with their favorite artists (Baym, 2012).

This ecosystem evolution leads to a more complex environment, in which music fans seldom *own* the music they listen to, as they would by purchasing records or digital audio files (Morris, 2015; Vicente, 2012). Big music catalogs are available for personal online streaming through multiple platforms, compelling record labels to adapt their target consumers from record buyers to the much broader group of potential digital streaming listeners (Gurgel, 2016). Music is now offered via digital platforms in what Burkart (2014) presents as the *Celestial Jukebox*. These platforms shift the music market from one that sells plastic products containing copyrighted recordings into the other that allows the listening of those recordings as a service, removing the ownership of the physical support from the equation (Anderson, 2013). These digital platforms are created in a way to enable unlimited scalability, and they become more profitable as they gather more subscribers (Vonderau, 2019).

At the same time, music fans gather in social platforms to share their own personal music collections through *playlists*, making the audience itself – whether this choice is filtered by algorithms or not – a major actor responsible for introducing new music genres and performers to their peers. The audience has long been taken into account when studying media and cultural businesses, especially when they are supported by advertising. Philip Napoli (2003) laid the groundwork for understanding the targeted audience as those who purchase the advertised products, and not as the ideal consumers of the provided content. In this case, content is used as a "bait" for promoting products and services. As micro-targeted advertising fuels the music streaming business, music becomes a means to an end in the business equation and not the other way around – musical tracks are now better classified as "content" according to Negus (2019), used for attracting users to generate clicks and tracking their preferences to provide accurate consumer targeting. Nieborg and Poell (2018) furthermore consider such content to be "contingent" to platforms' interests.

Platforms track users through *algorithmic individuation*, creating ever-changing *personas* based on their current and contextual musical preferences; their playlists; and, in services such as Spotify, their friends' preferences, in order to offer them the most possibly accurate ads (Prey, 2018). Although portable music with earphones has been around since the walkman (Hosokawa, 1984), by listening to music in the cloud users have their listening habits tracked and end up being commodified themselves, as Burkart (2013) puts: "The user-generated content that makes these services appealing to users and investors is unremunerated fan labor" (2013).

While the change from a business focused on selling records to a service-oriented one is not comprehensive, it is gradually becoming natural for music industry players to measure success according to digital ratings (such as *total monthly listeners*, *streaming counts*, and *video views* of individual tracks), bringing data into the musical equation.

This chapter drafts gatekeeping from a *datacratic* perspective. A *datacracy* is a social regime in which data and algorithms are influencers on how decisions are made, in such a way that algorithms are ranked alongside human experts, with similar magnitudes (Radfahrer, 2018). From this perspective, intertwined relationships among people, algorithms and databases, are increasingly being defined by the machine, which is developing a strong – and growing – influence upon social decision-making processes.

The music industry can be understood as a practical application of a *datacratic* regime in itself, for all major strategies – including marketing, investment, and even audience interest – are influenced by the industry numbers, and this creates an endless feedback loop. An artist's sales or *play counts*, when published, influence their future sales and plays by branding them as successful and interesting.

The reading of music sales data as marketing intelligence is not novel, nor a phenomenon of digitization. In 1991 Billboard already used SoundScan, a bar code-based system that tracked record sales in stores, and its results could unscrupulously determine the rise or fall of artists (McCourt & Rothenbuhler, 1997). This was a change to an industry previously based on "whatever sticks" (Drew, 2014), knowing few of the produced records would turn a profit, but reluctant to study how and why (Straw, 2001). The current access to data, however, is no longer analyzed by marketing executives, and is not gathered only in the final stage of selling the finished product. Programmers in large TV and radio stations rely on play rankings in order to validate their curatorship according to Playax, a Brazilian company that provides such services and is the primary data source to build this paper. Current data and feedback are available directly to artists themselves during their pre-production, as early as creating music and thinking on what subject to write about for a new unconceived album, as noted by artists interviewed by Nancy Baym about their relationship with fans through social media (2012).

The *phonograph effect* (Katz, 2010) proposes that musical creations are changed by the technologies involved in their making, reproducing, and distributing. Just as Katz demonstrates that the invention of the phonograph influenced music to be shorter in length, the rise of streaming services facilitates the release of singles and EPs instead of full albums and, even further, promoting and testing of music through social media and the feedback it generates changes the artistic outcome of the final recording. When analyzing taste profiles assigned to users based on their listening habits, Robert Prey (2017) wonders "how long until music is tailor-made to match these profiles?", questioning how platformization might affect the creative autonomy of musicians. Data-driven customization of culture is already taking place according to Nieborg and Poell (2018). Spotify pays royalties proportionally to track plays of 30 seconds or more, and insightful creatives have figured out that splitting music into shorter tracks generates more revenue. Ingham (2018) exposed a strategy in which playlists were filled with soundtrack-like tracks by seemingly unknown artists, lasting less than a minute, apparently made from scratch to fit specific playlist moods. This specific relationship of data with the artistic creation makes the music industry fit into the concept of a *datacracy*.

In the light of this discussion, we chose to trace the path of fast-rising artists through digital platforms, aiming to identify how different genres may have different relationships with the digital distribution of music to their audiences. Different genres have distinct audiences, as well as independent and *major* financed music are produced in dissimilar ways (Born, 1987; Negus, 2004). We hypothesize that genres will flow in dissimilar ways from production to listening.

This proposal may be framed as part of a broader perspective, towards setting a basis for understanding the overall impact of digital technology on the communication surrounding musical releases, and the understanding of music as content. Our study is focused on the Brazilian environment, but we believe it can be easily applied to other contexts, taking cultural and market differences into consideration. The model chosen was used as a guide to analyze streaming, broadcasting, and social media data from radio, YouTube, and Spotify provided by *Playax*, an audience analysis platform aimed at music industry professionals.[1]

On Brazilian music industry particularities and its genres

There has been a struggle concerning forces driven by creativity and commerce in the core of the music industry since the early 1960s, noticed by Simon Frith when he signals that "art discourses were beginning to be applied to 'commercial' sounds" (Frith, 1998). This is also pointed out by Hesmondhalgh and Baker, referring to the use of the term *cultural industries* due to the rising reverence of the concept of creativity in the same decade (Hesmondhalgh & Baker, 2011).

The concept of popular culture might be explained as a way of mediating class and group values in which genres in popular music bring forward the "issues thrown up by their commodification" (Frith, 1998, page 45). While each music genre brings a different set of values from the sociological point of view, and may be listened to by an audience of similar values, it might as well be consumed as a commodity through a different set of platforms, of which specific audiences are more familiar with. This will be the case with our selected Brazilian genres ahead, which are uniquely local, rooted in Brazilian history and drenched in social clashes.

Besides the adoption of genres as one of the axioms to understand popular music, this research also takes into consideration the characteristics implied by being released by independent labels and major record labels, representing creative and commercial values, respectively. Keith Negus makes a compelling argument that independent labels, in comparison with major record labels, have more awareness of new trends and are freer from big commercial ties to pursue them, therefore being responsible for releasing to the audiences new music that could not even be recorded if it relied only on major record labels. At the same time, the author makes a reminder not to over-idealize independent labels, since they might act as talent scouts for major record labels, which release following projects of "pre-tested" independents (Negus, 1996).

On the other hand, when major record labels take successful independents and turn into commercial hits, they keep their status as culture setters, as described by Patrick Burkart when analyzing the new digital music environment in the beginning of the transition to a streaming model: "The new oligopoly in the music business continues to exert anti-competitiveness throughout its traditional distribution bottlenecks" (Burkart, 2005). We propose that such bottlenecks can be identified by applying the *digital music gatekeeping* model into music communication, and we begin by testing it in the Brazilian music industry.

While the global music industry grew 9.7% in revenue in 2018, Brazilian music industry stood out with a 15.4% growth, becoming the largest market for recorded music in Latin America and occupying the tenth place globally (IFPI, 2019). Brazilian numbers were driven mainly by the adoption of streaming platforms, which were responsible for 69.5% of total revenue in 2018 and represented a 46% growth year over year (ProMusica, 2019).

Brazil was noted as one of the fastest growing markets of the year by IFPI's *Digital Music Report* (2019). In addition to the noticeable growth of the music market in Brazil, comes the particularity of the domestic genres – most of the music listened to in Brazil is also made in Brazil (Wikstrom, 2009), and the analysis of the 200 most popular tracks played in Brazilian streaming platforms in 2018 reveals that 84% of them are performed by local artists (ProMusica, 2019). When the same list of tracks is aggregated by its genres, *Sertanejo* shows up as the most popular genre with 44% of them, while *Brazilian Funk* is found in second place with 35%. The two genres are typically Brazilian and will be the context of selection for our five cases.

Understanding *Brazilian Funk* and *Sertanejo*

We have chosen to work with five artists from two of the most popular musical genres in contemporary Brazil, *Brazilian Funk* and *Sertanejo*, which are also typical Brazilian music. What is called *Brazilian Funk* is very different from what is known internationally as *Funk*, the genre originated in the African-American communities of the USA in the 1960s, popularized by James Brown. *Brazilian Funk*[2] bears no relationship with the international Funk apart from its name, coined in popular parties happening in Rio de Janeiro's favelas in the 1970s, in which international Funk and soul music would also be played (such parties are still going strong today, despite the fact that the music played there changed over time). Brazilian Funk parties and their organizers are widely known for their low-cost tickets, ubiquitous cheap alcoholic beverages, large and loud sound equipment, having their own media outlets (radio, TV, and Internet) and a star-system of DJs[3] and MCs.[4] What is called *Brazilian Funk* music is usually derived from one of these three major genres: rap, mash-ups created by DJs, and short melodies with double meaning lyrics (Sá & Miranda, 2011).

In the 1990s and 2000s, it was already clear to the *Brazilian Funk* business that having a hit song did not mean boosting their record sales – especially due to the widespread piracy in Brazilian popular music.[5] Having a hit song usually meant boosting parties' attendance profiting from ticket and beverage sales (Sá & Miranda, 2011).

The company *Furacão 2000* was a strong representative of the Rio de Janeiro branch of *Brazilian Funk* by organizing parties, releasing artists, and managing media professionals to promote them as early as the 1970s. These days, other producers such as *KondZilla* from São Paulo, a video producer and record label, whose YouTube channel is one of the 10 most viewed in the world,[6] are breaking into stardom as well.

Before diving into the case study of five *Sertanejo* and *Brazilian Funk* artists, we will discuss how we can do so by understanding how their communication is spread in order to make them known by their potential audience.

Gatekeeping, datacracy, and platformization

Gatekeeping theory was first presented by Kurt Lewin in 1943 (Lewin, 2014). Originally envisioned under a psychology context, the term described the decision-making process of housewives when purchasing groceries. One of the first implementations of this theory in a communications context was by David M. White (1950), reporting an editor's choice process when selecting the news to be published in a newspaper. Since then, there has been a lot of debate over the depth of the gatekeeping process (Bass, 1969; Bruns, 2003; Singer, 2014).

When discussing cultural products, Paul M. Hirsch supports Brown's broader chain of gatekeepers, by presenting the record label as the first gatekeeper of music – the one which decides what kind of music will be recorded (Hirsch, 1972). Hirsch analyzes music gatekeepers in a context of narrower selections of available music, provided by the workflow of major record labels' releases for a physical goods market. These past two decades are typical for broader selections of music made available through digital catalogs; however, the clustering of decisions by major record companies and mass media outlets is significant still.

Long tail is Chris Anderson's (2008) designation of the large amount of cultural products that sell in small quantities and therefore were not profitable for large stores to keep in stock in the 20th century. Hirsch (1972) mentions these logistic limitations to less successful music albums. These products, however, with the possibility of being sold remotely and free of geographic boundaries in online shops, are summed up to a large amount of sales and become profitable, according to Anderson (2008). When listening to music through platforms, the audience might encounter tracks that they might not buy, for a number of reasons including that they might not even have heard of such artist. Platformization enables the distribution of *long-tail* selections of independent music and the casual listening of them by those who might or might not become their fans. Streaming platforms create narrow niches of service with extremely curated content according to Morris and Powers (2015), in a way similar to cable TV channels, however generating a large amount of data about users' listening habits. Robert Prey (2016) proposes that this *datafication of listening* gets in the way of that perceived freedom of finding music, since the platforms' *taste profiles* create *filter bubbles* (Pariser, 2011) around the listeners. Playlists are at the center of these bubbles, as Negus (2019) notes: "In an age of abundance the curator becomes more significant than the creator. The playlist becomes more culturally and commercially important than the idea of the album as artistic statement and commodity". We will propose further along, in our *digital music gatekeeping* model, that these filter bubbles act as gatekeepers as well, by selecting which tracks go through their gates.

Datacracies can be understood as new social regimes in which data plays the most important role, defining and ruling social interactions (Radfahrer, 2018). We draw from the concept of *datafication* in order to understand how the world becomes quantified in every aspect, through the massive gathering of data from user interaction on digital platforms, which led to the transformation of technology companies business models from technology oriented to data oriented (Dijck, Poell & Waal, 2018). *Datacracy* is an evolution of Neils Postman's *Technopoly* (1993), in which technological tools are used to provide direction and purpose for society and individuals: "a society in which technology is deified. The culture seeks its authorisation in technology, finds its satisfactions in technology and takes its orders from technology".

Datacracy takes on a broader view than *data colonialism*, since it is not restricted to cultural influence and/or domination. When focusing on Brazilian music landscape, it is noticeable that local artists and cultural demands play a significant part in music production, from creation to distribution, and we draw on the *datacracy* model to focus on the consumption of Brazilian music by locals instead of the *data colonialism* approach on "predatory extractive" quantification methods (Couldry & Meijas, 2018). In the music realm, it has a *dataveillance* (van Dijck, 2014) view, in which its actors swap their roles, albeit this practice monitors and collects online data and metadata through social networks and online platforms, it is not used as surveillance, but as a marketing-based creative tool, which tends to provide consumers with exactly the kind of entertainment they expect.

The music industry has a history of being influenced by market demands, which tend to have a strong impact on creative choices. This information, however, was gathered by research market services, and mediated by editors and labels, in a process that was far from direct. Social media's rising popularity allied with the growth of on-demand streaming music services provided artists with direct access to live consumer data through specialized dashboards such as *Spotify for Artists* and *YouTube for Artists*, having detailed fan interaction live data that allows detailed marketing strategies for artists to reach and expand their target markets, demonstrating the direct impact of every creative decision, a process that tends to enhance the influence of market data on creative output.

This scenario fits a *datacracy* context, in which Big Data systems tend to have an increasing influence over the general public. Algorithms, in a *datacracy*, are not simply seen as equations – instead, they match Nick Seaver's (2017) approach as "heterogeneous and diffuse sociotechnical systems, rather than rigidly constrained and procedural formulas". We rely on the concept of *datacracy* as the foundation of our *digital music gatekeeping* model, suggesting that the music industry is in itself a *datacracy*, governed by data while flowing through the platforms.

Patrik Wikström's *Audience-Media Engine* is also taken into account in our proposed model, by including *audience approval* and *audience action* as variables connected to *audience reach* and *media presence*; all of which will together work towards or against the promotion of an artist. While the *media presence* and its resulting *audience reach* might be generated by standard gatekeepers like mass media outlets, *audience approval*, and *action* are generated by the listeners themselves (Wikström, 2009). By including them in the loop with the first two, Wikström brings the *Audience* to the same level of importance as traditional *Media* in his *Engine*, and as it loops over and over, the measurements of these audience reactions can become just as important as the reactions themselves. Music industry data becomes a sort of content in itself, and in its *datacratic*-oriented regime, it takes part in determining whether a given artists' loop will go forward or backward. For this reason, artists are increasingly concerned not only with selling, but also with rising in the charts to tell their audience they are indeed well-sold.[7] We propose that the audience factor might be more relevant than the major media outlets in the current system for some genres, and we will return to this when comparing our two selected genres.

Shoemaker and Vos (2009) have also taken the audience into account in their *gatekeeping Theory*, even if secondary. As news items are offered by major media on digital outlets, they are ranked according to their popularity among readers. This secondary filter is integrated into their gatekeeping theory in order to explain the reach gap between different news items. The researchers pointed out that "having the capability to access huge amounts of foreign news on the internet does not mean that individuals will read it" (Shoemaker et al., 2010, p. 67).

Wallace's (2017) analysis of digital journalism in the news media landscape takes into account the rise of algorithms and platforms, introducing new variables to the gatekeeping process, while also demanding a revision of the classical model. Instead of only working with primary and secondary stages of gatekeeping, Wallace proposes a new model for news dissemination made of three consecutive stages in a cyclical iteration: input, throughput, and output. His model also suggests four types of gatekeepers acting through the stages as agents of information flow: *journalists*, *algorithms*, *strategic professionals*, and *individual amateurs*.

In the Wallace model, *input* stages acknowledge that agents have different levels of access to information. Journalists, for example, can use an organized source structure provided by a news agency, while amateurs might rely on personal information networks. *Throughput* stages consider each information agent's selection criteria. Journalists, for example, tend to check facts and sources aiming trustworthiness, while newsgathering algorithms such as *Google News* redistribute already published news items favoring popularity and other behavioral metrics to increase audience consumption. The *output* stage is defined by the available choices of platforms for content to be published. Once three stages are completed, the resulting publication might be used as an information source for a new cycle.

Wallace's multi-node model is consistent with the shift from two-sided market structures to "complex multisided platform configurations" proposed by Nieborg and Poell's (2018) work on platformization. As essential intermediaries for publishing and distributing content in the digital age, platforms in the Wallace model are considered scenarios containing rules for gatekeeper interactivity on the third stage. Each platform acts in a particular way, offering advantages or disadvantages to information flow, depending on their structure, interface, and user interaction. In regard to these characteristics, Wallace's model proposes two sorts of platforms, according to their gatekeeping mechanisms, them being *centralized* or *decentralized*. In centralized platforms, information comes from fewer and more controlled sources (such as newspapers, magazines, news agencies, and newsgathering algorithms such as Google News), which supposedly can vouch for their output and reach the audience more easily. In decentralized platforms, anyone is able to publish content (such as social media, blogs, and forums), and these items demand a lot of public interaction to avoid oblivion.

Following Wallace's model, algorithms are applied in two different ways. As gatekeeper agents, they harvest and reorganize content previously published by other agents, and output them on centralized platforms. Google News is an example of a newsgathering algorithm in which content relevance is mathematically calculated by readers' responses in a massive scale. For our proposal, we understand this type of iteration as the one from algorithm-curated playlists such as the automatic and individual *Discover Weekly* from Spotify. Each user is given a different selection based on their listening history and current trends.

However, algorithms running in social media are understood as structural features of decentralized platforms that build relevance over networked content, and not as gatekeepers themselves. This is the case of the Facebook newsfeed algorithm that controls content availability according to peers interactivity. In our proposal, this type of algorithm would be a mediator of the nodes of decentralized platforms.

A *digital music gatekeeping* proposition

Given that music business decisions are increasingly being made following data measurements, we believe that an analysis and better understanding of these processes is essential. Our intent is to make use of an updated framework, grounded on Wallace's *digital gatekeeping* model and the concept of *datacracy*, while also relying on Wikström's *Audience-Media Engine*'s loops, Sauerberg's (2009) *Gutenberg parenthesis*, and Eli Pariser's (2011) *filter bubbles*.

By taking both human agents and algorithms into consideration, we can understand how they can act together with traditional mass media to influence listener choices.

We propose to interpret music as content in this model, instead of Hirsch's (1972) description of them as cultural products, which would make them nonmaterial goods. This is supported by Negus's (2019) claim that tracks become so when used as a means to generate ad clicks. The audience has – in many cases reluctantly – relinquished ownership of the music products and is now subject to hiring the ability to listen to whatever music is available on a given platform for a finite duration of time, with no guarantee of keeping a specific artist's music in their collections. If an artist decides to leave a platform, such as Taylor Swift has done after disagreeing with Spotify's remuneration system (MBW, 2014), the artist's fans instantly lose access to their music, even if saved and downloaded in their Spotify collections.

Parallel to that, we suggest to consider all the systems involved in promoting a track as equally capable of influence in a never-ending cycle, constantly fed into itself. Hirsch (1972) delineates gatekeepers in an aura of a few centralized and unreachable mass media outlets, consistent with the inside of the *Gutenberg parenthesis*. Our proposal is to understand, based on Wallace's model, that any and all nodes in this intricate web are also gatekeepers, potentially capable of as much influence as any other depending on the context – which includes the music genre. The key point in this detachment among all selectors is what kind of systems tracks are published *into*, whether they are centralized or decentralized.

Major record labels have well-equipped marketing and promotion departments as well as larger budgets, features which make them more likely to reach their audience through major media outlets, which is consistent with the bottleneck for major label releases suggested by Burkart (2005). Independent artists, however, mostly depend on aggregators to make their music tracks available in platforms. Once there, their tracks still depend on many interactions to be noticed by recommendation algorithms, human curatorship, and finally their end users, in a decentralized manner.

Given their massive audience, we believe that there are centralized and decentralized agents contained in both Spotify and YouTube. While larger channels and curated playlists are centralized, their low-reach user-generated channels and playlists are decentralized. This means that we can no longer consider these platforms as uniquely centralized or decentralized, since they contain both centralized and decentralized processes in themselves. Due to this recursive proposition of subcategorization of agents inside the same given platform, we will further address to the agents involved in letting music pass or not as centralized and decentralized *gatekeepers* instead of platforms. We propose the following ones to be included in the *digital music gatekeeping* model (Figure 6.1):

- **Centralized gatekeepers**, boosting the information through clear strategies, facilitated by filtering algorithms:
  - *Major record labels*, representing a somewhat official source of music tracks;
  - *Radio* and other official media outlets;
  - *Curated playlists*[8] and *large YouTube channels*, which are human-updated sources for distribution of new music to previously targeted audiences;
  - *Algorithm-curated playlists*, which are non-human playlists, ran by algorithms, such as Spotify's "Discover Weekly" and "Release Radar". These are automatically assembled for each individual user, based on listening history.
- **Decentralized gatekeepers**, influencing information spread through uneven paths, mediated by filtering algorithms:
  - Independent artists, promoting their own music through aggregators and their own media;

- *Strategic professionals*, being these the ones hired to promote certain contents throughout the media, such as public relations professionals, press agents, independent radio promoters, advertising in social media and platforms, paid placing in strategic spots; and also practices such as of *payola*, which is not illegal in Brazil, although it is frowned upon and mostly carried out in non-literal ways such as cultural support for radio stations and gifts;[9]
- *Individual amateurs*, which are non-corporate content producers including bloggers, YouTubers, social media celebrities as content prosumers and sharers themselves, and included are fan or artist-based strategies to boost playlists and independent lower-reach marketing efforts;
- The *Audience* in itself, as unpretentious influencers of their own smaller bubbles.

We intentionally did not include Spotify, YouTube, social media such as Facebook, Instagram and Twitter, and other smaller or alternative platforms such as Soundcloud and

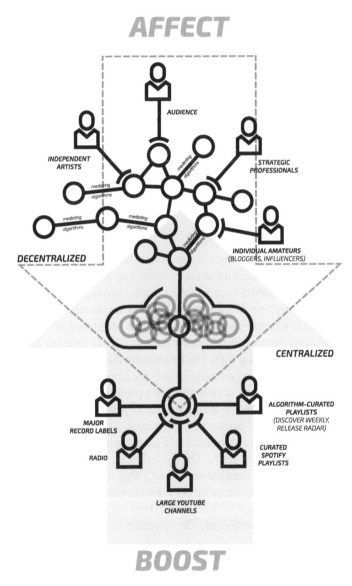

Figure 6.1 Suggested model for digital music gatekeeping.

Bandcamp in this model, understanding that they are platforms and not agents themselves, platforms being the scenario through which agents flow. Throughout each of these platforms, we shall encounter different centralized and decentralized agents. Billboard's YouTube channel would be considered a centralized agent in this model, opposed to an independent and little known vlogger using the same platform to promote his musical curatorship, and this vlogger would be considered a decentralized agent.

We also work with two different kinds of algorithms that must be clearly differentiated. Both kinds are human-designed equations that follow the procedures specified by their creators, letting content deemed "interesting" go through and vice versa, according to their rules (Beer, 2017). Algorithm-created playlists such as Spotify's *Discover Weekly*, the same as newsgathering algorithms like Google News, are agents in themselves, creating new curated content (playlists) based on a set of rules. Filtering algorithms are those which mediate the centralized and decentralized agents, and its rules of allowing the content to go through or not are central in determining if an agent will be considered centralized or not. Centralized agents are those which benefit from filtering and are able to boost their content through a larger audience, while decentralized ones are those which spread their content through uneven paths and have to survive through filtering to reach their destination.

We further suggest that these processes will vary throughout genres. While *Brazilian Funk* is notable for its audience strength and renegating mass media for its promotion, *Sertanejo* is known for having large and mostly private investors, able to place unknown artists in traditional mass media such as top radio stations through strategies similar to payola. This would suggest that *Brazilian Funk* artists would begin their path through decentralized agents, and *Sertanejo* ones backed by centralized media. We will further study the differences between these processes in individual genres through the gathered play count data ahead.

Methodology

Playax was chosen as the source for tracking audience data about Brazilian music industry. It is a service that creates a behavioral database by assessing an artist's performance in several platforms, from traditional media broadcasts to digital rental and streaming services, and provides aggregated statistics out of the data gathered. Our initial dataset included Brazilian artists with their respective music tracks and play counts originated on Spotify, YouTube, and radio platforms over a one-year period, from November 2017 to October 2018.[10] The standard metric used for comparison among these platforms were *daily counts* of played music tracks.

Our methodology was based on audience behavior towards artists separated by genres, looking consistent growth in music distribution as a premise to identify centralized and decentralized gatekeepers that might be acting on platforms. For example, if there's a growth trend associated with a group of artists of a particular genre originated in a specific platform, that fact was considered as an evidence of a gatekeeping process happening within the platform and associated with that genre. On the other hand, isolated spikes were signaled in the analysis due to the lack of consistency growth that might be explained either by paid media usage to promote new releases or by occasional mass media appearances.

The initial dataset was processed to choose 50 artists with most consistent and significant growth in the 12-month period analyzed. The selection process to achieve that result is described below:

1. *Playax Ranking* is a metric provided by *Playax* which sorts all artists by overall audience in every period. We calculated the year average ranking of each artist from November 2017 to October

2018 in order to assure that growth was consistent, lessening the impact of growth spikes caused by occasional events.

2. For each artist, the year average was compared by subtraction to the artist ranking on November 2017, resulting in a range that was used to identify growth trends over the year. A larger positive range was an indication that the artist had gained popularity during that period.

3. After that, we filtered only artists who were among the thousand most popular, sorted by the year average ranking. This step in the process assured the removal of artists with large positive growth that are deep down in the ranking (such as raising 100,000% from 3 to 3,000 streams).

4. Finally, after applying the filter, we sorted the artists by their growth trend found in the second step and selected the 50 steeper growth curves.

When analyzing the final group of 50 artists, *Brazilian Funk* and *Sertanejo* genres stand out, not only by the amount of artists (11 on *Brazilian Funk* and 7 on *Sertanejo*), but more specifically because of the similar paths followed by their curves. From the original sample, we chose to select the five steepest and idiosyncratic curves of both genres for a more dedicated analysis. Because they are so different in audience and promotion strategy, we believe these two genres will significantly draw visible differences in our *digital music gatekeeping* model.

Brazilian Funk artists

KondZilla, an original music video producer YouTube channel, appears to be a centralized gatekeeper in itself. Their reach can be noticed when analyzing the data of artist *MC Loma & as Gêmeas da Lacração*. In January 20, 2018, the artist published an amateur self-made music video for the track "Envolvimento".[11] From then to the first days of February, they rose sixfold, from an average of 500,000 to about 3 million daily views. However, it was when the same song had a video produced by KondZilla and released in their channel, they peaked 11.8 million views in a single day,[12] the first day after the video was published. This one by KondZilla reached 233.6 million views by November 2018.[13] But more than just pointing to their peaks, Playax's dataset revealed how their boost in YouTube views reflected in more Spotify and radio streams in the following months (Figure 6.2).

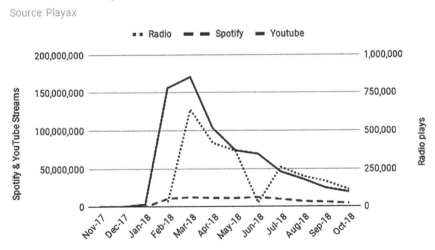

Figure 6.2 Playax data of artist MC Loma & as gêmeas da lacração.

From this data, we recognize a path through the gatekeeping model which starts with an independent artist publishing a track via YouTube, this track being repeatedly shared by the audience (Figure 6.3). Later on the track is published at KondZilla's YouTube channel, a centralized gatekeeper, leading to more sharing and the addition of the track to Spotify curated playlists such as "Segue o baile",[14] "Funk Hits",[15] and "Funk 2018 – Melhores funks

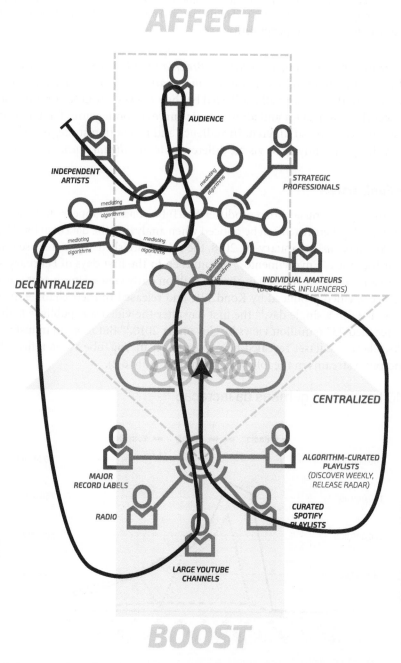

Figure 6.3 Suggested gatekeeping cycles of MC Loma & as gêmeas da lacração.

2018"),[16] parallel to the automatic recommendation of the track through algorithm-curated playlists and the extensive sharing by individual listeners.

The same rising pattern through *KondZilla* is seen in *MC Dede*'s data, which holds a similar path through the musical gatekeeping model. On April 21, 2018, when a video was published at the KondZilla channel,[17] their daily views peaked to 3.5 million, a huge growth from the day before, an audience of barely 200,000.[18] In the following months, their Spotify presence rises significantly and, although their YouTube presence decreases, their Spotify audience keeps steady throughout a few more months (Figure 6.4).

Sertanejo artists

While *KondZilla* might be one of the major YouTube channels centralized gatekeepers when it comes to *Brazilian Funk* music, it does not necessarily hold its influence for other music genres. The popular *Sertanejo* genre calls for new insights. Despite being originated from the Brazilian heartland (contrasting with the metropolitan sprung Funk) and sometimes related to Country music genre in the USA, Brazilian *Sertanejo* has a history of its own.

Sertanejo was originally the folk music of Brazilian rural regions and was quite different from what is heard today. It became marginalized during the second half of the 20th century and was cast aside by the elites as a lesser musical style (Alonso, 2011). By the end of the century, it was rebranded as a new commercial genre, institutionalized and called *Sertanejo Universitário*, meaning it is predominantly made and consumed by college students (denoting elites). Although still being made by mostly duo artists, their lyrics went from complaining about cheating spouses to a constant elevation of love itself and the commemoration of the end of bad relationships. Although *Sertanejo* has risen to the top of most charts above pop music, their artists claim to be different, because they are mostly only affiliated to record labels for distribution, and not for comprehensive deals (Alonso, 2012).

Sertanejo artists' data indicate that they might follow a different path from the one taken by *Brazilian Funk*. We believe that one of their main gatekeeping mechanisms is the Spotify

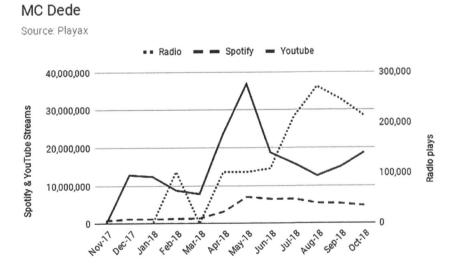

Figure 6.4 Playax data of artist MC Dede.

playlist, opening the gates to being discovered and listened in the platform. Artists such as *Lucca & Mateus* begin their trajectory by rising significantly on Spotify, then spreading to YouTube and radio on the following months. Radio plays were also boosted at the same time, however in a much smaller rate, as shown in Figure 6.5 (please note that radio plays refer to the right-side axis, in a different scale, in order to reflect the fact that radio plays reach a large number of listeners, while YouTube and Spotify plays reach mostly individual ones).

Although it is not possible to spot which specific playlists this artist was added to in this specific time frame, due to the platform's analytic restrictions, Spotify states in the artist's profile that they were discovered on playlists such as "Sertanejo no trabalho",[19] "Sertanejo universitário 2018",[20] and "Viagem sertaneja".[21] This artist seems to have steered through the gatekeeping model in a different way. While *Brazilian Funk* artists started on decentralized gatekeeping agents and were then re-released on KondZilla's large and centralized YouTube channel, Lucca & Mateus were first popularized on Spotify, suggesting a first point of contact with the audience through centralized outlets such as human-curated playlists. In other words, we propose that agents which could generally be considered as decentralized such as large YouTube channels and curated Spotify playlists are in fact centralizing agents due to their facilitation of discovery through the platform filtering. Artist Hugo & Guilherme also seem to reflect Spotify's growth into YouTube, suggesting that their first point of discovery was on Spotify (Figure 6.6).

However, both Lucca & Mateus and Hugo & Guilherme show steep peaks in YouTube and radio. These quickly rise and fall in a single month, suggesting they might be exceptional situations. Radio and YouTube peaks seem to be exceptions, which may be the quick rise and fall of a hit, but may also be generated by advertising efforts in that month, more specifically YouTube advertising and radio payola in this case. This is consistent with a large peak that does not hold itself in the following months (or at least some of it, as happened with the studied *Brazilian Funk* artists). The strategic investment in artist Lucca & Mateus appears to generate quick vanishing peaks in their visibility; however, it does not seem to raise their organic YouTube and radio plays significantly, given that the play counts drop back down right after the month of the boost.

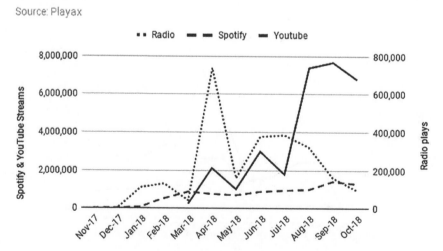

Figure 6.5 Playax data of artist duo Lucca & Mateus.

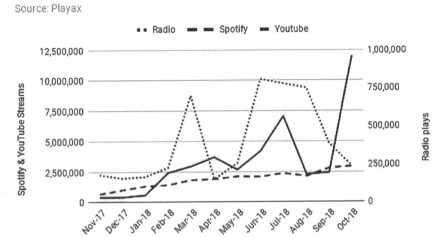

Figure 6.6 Playax data of artist duo Hugo & Guilherme.

While we are limited to the public data of views, streams, and radio plays, we are not able to test the hypothesis of them being generated by advertising and payola without inside data on investments. However, we can indicate that these peaks are much steeper than the ones in the analyzed *Brazilian Funk* artists.

These peaks become even clearer to point out in artist duo *Otavio Augusto & Gabriel*, whose chart is mainly made of peaks that are not sustained in the following months (Figure 6.8). Also note that radio plays in Figure 6.8 refer to the right-side axis, which now uses the exact same scale of the Spotify and YouTube left-side axis.[22]

Spotify streaming numbers, however, do not show peaks like these for the analyzed artists, sustaining the company's official declaration which states that "you cannot buy your way into a Spotify playlist".[23] Third-party curated playlists, however, do not always share this same directive, and deals for placing tracks in them might be offered. We infer that this music might have been recommended via algorithm-curated playlists such as *Discover Weekly* to their target users, as well as included in human-curated playlists (whether hired or organic), both centralized agents of the gatekeeping process, in accordance with Robert Prey's (2017) *datafication of listening*. On our goal of identifying which gatekeepers influence other gatekeepers to recommend a musical track, the Playax data suggests that other efforts on YouTube and radio were not significantly influential to the growth of their listener base on Spotify.

Closing remarks

Our journey through emerging Brazilian artists shows that there is not a fixed or standard model for promoting music anymore. It might have been the case in the 20th century analyzed by Hirsch (1972), when major labels and major gatekeepers boosted tracks around to the audience through their less resistant pathways. From the optics of a *datacracy* (Radfahrer, 2018), as the data and its processors, the algorithms, become active stakeholders in the gatekeeping path and artistic decisions get made because of such data; we present this *digital music gatekeeping*

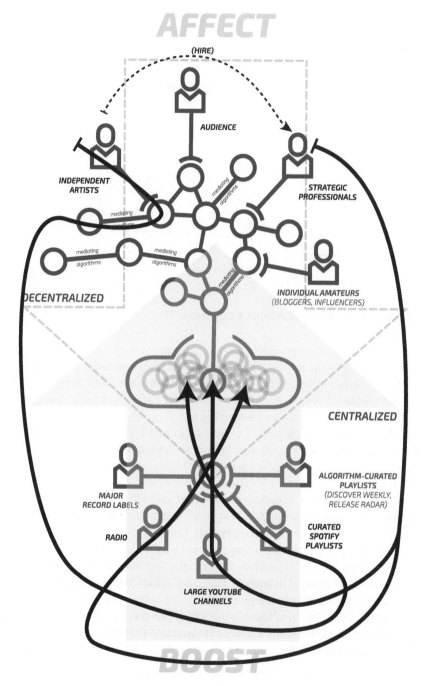

Figure 6.7 Suggested gatekeeping path of Lucca & Matheus.

study, supported by Wallace's inclusion of decentralized platforms in his *digital gatekeeping theory* (Wallace, 2007), Wikström's *Audience-Media Engine* (Wikström, 2009), and platformization (Dijck et al., 2019; Nieborg & Poell, 2018) theories along with current music business and digitization studies (Negus, 2019; Prey, 2018; Leyshon, 2014; Vonderau, 2019).

Otávio Augusto & Gabriel

Source: Playax

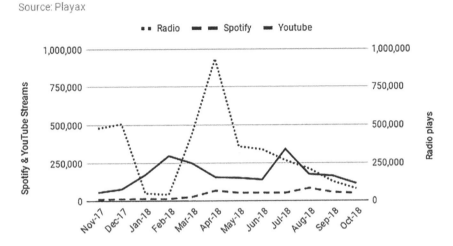

Figure 6.8 Playax data of artist duo Otavio Augusto & Gabriel.

This study attempts to propose updates and extensions to the Wallace *digital gatekeeping* model by applying it into music. Despite being the case related here a very specific case study with a few artists in the very particular Brazilian music genres, we strongly believe that it can be spread to other music genres and cultural contexts, provided that cultural and social attributes are taken into consideration along with similar datasets.

This research is still in its beginning and demands further explorations to be fully understood. But the unmistakable correlations between YouTube channels activities and radio track plays (from the data gathered by the *KondZilla* influence on the *MC Loma & as Gêmeas Lacração* video), and between Spotify curated playlists and radio track plays (from the data gathered by some *Sertanejo* artists) show that these digital music streaming actors play a strong role as music gatekeepers in this new digital, social media environment. It also raises the question to study whether strategic professionals and advertising are able to influence their results in the same way they were in the 20th century. Despite artists from different musical genres display different patterns in their gatekeeping processes, we strongly believe that contemporary success strategies depend heavily on gatekeepers acting on digital streaming platforms, in genres as diverse in their musical style and audience as *Brazilian Funk* and *Sertanejo*.

We hope this research contributes with factual data to the present-day debate on the importance and influence of digital platforms in audience choices, encouraging further studies on the field.

Notes

1. Available at www.playax.com. Accessed in 30 Nov 2018.
2. *Brazilian Funk* is mostly known in Brazil as "Funk Carioca", which means "Funk from Rio de Janeiro", or simply "Funk". However, it has since been produced in many other regions of Brazil and it requires distinction from international Funk, so it has been categorized in our original database as *Brazilian Funk*.
3. DJ: Disk Jockey, those who play and combine music for people to dance (Sá & Miranda, 2011).

4.   MCs are the Masters of Ceremonies, who write and rap music, just as hip hop MCs (Sá & Miranda, 2011).

5.   Street vendors were common to have CD & DVD stands, offering pirate versions of hit records and their own compilations.

6.   According to YouTube data consolidated by Statista. Available at www.statista.com/statistics/373729/most-viewed-youtube-channels/, accessed in 30 Nov 2018. KondZilla had 27 billion views and 54 million subscribers on its YouTube Channel in 24 Dec 2019. Available at www.youtube.com/user/CanalKondZilla.

7.   The industry seems to be aware of how it is being run in loops by its own numbers, up to the point in which artist DJ Khaled has threatened to sue Billboard for being charted number two in a week in which his own math would make him number one: "'Monster' US charts fallout: is DJ Khaled about to sue Billboard after missing out on no.1?", in Music Business Worldwide, by Tim Ingham. Available at www.musicbusinessworldwide.com/dj-khaled-reportedly-planning-to-sue-billboard-after-missing-out-on-no-1-chart-spot/.

8.   Including all kinds of curators, be them music critics, journalists, fans, or even brands which might use those playlists to set the mood for their brand image (Drew, 2005).

9.   Revealed by André Midani, main executive of the 1960–1990s Brazilian music industry, interview by Pedro Alexandro Sanches. Available at www1.folha.uol.com.br/folha/ilustrada/ult90u33266.shtml. Accessed in 30 Nov 2018.

10. In order to foster future debates, we enclose the Playax dataset used for this research (in .csv format) as an appendix to this paper, hoping that it proves itself useful as a testing ground.

11. This video totaled 47.5 million views by November 2018. Available at https://youtu.be/pOpyq-T4fnQ. Accessed in 30 Nov 2018.

12. Daily view count obtained from YouTube's own charts. Available at https://charts.youtube.com/artist/%2Fg%2F11f3_q6my0?date_end=2018-11-27T00%3A00%3A00Z. Accessed in 30 Nov 2018.

13. Available at www.youtube.com/watch?v=lgJOJAmXlBw. Accessed in 30 Nov 2018.

14. "Segue o baile", official Spotify playlist with 431,000 followers. Available at https://open.spotify.com/user/spotify/playlist/37i9dQZF1DWWmaszSfZpom?si=M8xJlcSeTZy5D2b-N7wokw. Accessed in 30 Nov 2018.

15. "Funk hits", official Spotify playlist with 2 million followers. Available at https://open.spotify.com/user/spotify/playlist/37i9dQZF1DWTkIwO2HDifB?si=5GohUbCfS26QAekeRcx50w. Accessed in 30 Nov 2018.

16. "Funk 2018 - Melhores funks 2018", Spotify playlist curated by redmusiccompany, 712,000 followers. Available at https://open.spotify.com/user/redmusiccompany/playlist/0WrqPR7s1X-73LBJiW2eRWC?si=aSyO4ndPSsSwt2pb89ac6Q. Accessed in 30 Nov 2018.

17. Available at https://youtu.be/2XHgXfmfsZU. Accessed in 30 Nov 2018.

18. Daily view count obtained from YouTube's own charts. Available at https://charts.youtube.com/artist/%2Fg%2F11b7_v3lwg?date_end=2018-11-27T00%3A00%3A00Z. Accessed in 30 Nov 2018.

19. "Sertanejo no trabalho". Official Spotify playlist, 330,000 followers. Available at https://open.spotify.com/user/spotify/playlist/37i9dQZF1DWUkWvWISwtjS?si=F2zomuFpRnWOa5WzU7T-KDA. Accessed in 30 Nov 2018.

20. "Sertanejo universitário 2018". Spotify playlist curated by Vinicius Menegola, 99,000 followers. Available at https://open.spotify.com/user/vini_kam/playlist/4C7la5DHqabGdAa-J3615I5?si=U06dDutdRcSdKt7TY0WJbg. Accessed in 30 Nov 2018.

21. "Viagem sertaneja". Official Spotify playlist, (Spotify, 25,000 followers). Available at https://open.spotify.com/user/spotify/playlist/37i9dQZF1DWUct0XKgeOXv?si=rxBa1mZwTwuGTqTcFy-W7mg. Accessed in 30 Nov 2018.

22. In this case, we used the same scale on both sides to keep the figure understandable. If we were to apply similar proportions as the other figures, Spotify and YouTube entries would be rendered almost flat at the bottom and would not be discernible for analysis.

23. Said by executive Meg Tarquinio, 36 seconds into the institutional video on "How to get playlisted". Available at Spotify for Artist's Playlist tutorial, at https://artists.spotify.com/videos/the-game-plan/how-to-get-playlisted. Accessed in 30 Nov 2018.

References

Alonso, G. (2001). *Cowboys do Asfalto: Música sertaneja e modernização brasileira*. (Doctoral dissertation). Universidade Federal Fluminense, Niterói, BR.

Alonso, G. (2012). O sertão vai à faculdade: o Sertanejo Universitário e o Brasil dos anos 2000. *Revista Perspectiva Histórica*, nº 2, p. 99–111, jan-jun 2012. Retrieved from http://perspectivahistorica.com.br/revistas/1434420384.pdf

Anderson, C. (2008). *The Long Tail: Why the Future of Business Is Selling Less of More*. New York, US: Hyperion Books.

Anderson, T. J. (2014). *Popular Music in a Digital Music Economy: Problems and Practices for an Emerging Service Industry*. In Routledge Research in Music. doi:10.4324/9781315850948

Bass, A. Z. (1969). Refining the "Gatekeeper" Concept: A UN Radio Case Study. *Journalism Quarterly*, 46(1), 69–72. doi:10.1177/107769906904600110

Baym, N. K. (2012). Fans or Friends?: Seeing Social Media Audiences as Musicians Do. *Participations*, 9(2), 286–316.

Baym, N. K., & Ledbetter, A. (2009). Tunes that Bind? *Information, Communication and Society*, 12(3), 408–427. doi:10.1080/13691180802635430

Beer, D. (2017). The Social Power of Algorithms. *Information Communication and Society*, 20(1), 1–13. doi:10.1080/1369118X.2016.1216147

Born, G. (1987). On Modern Music Culture: Shock, Pop and Synthesis. *New Formations*, 2(2), 51–78.

Born, G. (2005). Digitising Democracy. *Political Quarterly*, 76(SUPPL. 1), 102–123.

Bro, P., & Wallberg, F. (2014). Digital Gatekeeping. *Digital Journalism*, 2(3), 446–454. doi:10.1080/21670811.2014.895507

Bruns, A. (2003). Gatewatching, Not Gatekeeping: Collaborative Online News. *Media International Australia*, 107, 31–44.

Burkart, P. (2005). Loose Integration in the Popular Music Industry. *Popular Music and Society*, 28(4), 489–500. doi:10.1080/03007760500159013

Burkart, P. (2010). *Music and Cyberliberties*. Middletown, CT: Wesleyan University Press.

Burkart, P. (2014). Music in the Cloud and the Digital Sublime. *Popular Music and Society*, 37(4), 393–407. doi:10.1080/03007766.2013.810853

Burkart, P., & Leijonhufvud, S. (2019). The Spotification of Public Service Media. *Information Society*, 35(4), 173–183. doi:10.1080/01972243.2019.1613706

Couldry, N., & Mejias, U. (2018). Data Colonialism: Rethinking Big Data's Relation to the Contemporary Subject, *Television and New Media*, 20(4), 336–349. https://doi.org/10.1177/1527476418796632.

De Marchi, L. (2018). Como os algoritmos do YouTube calculam valor? Uma análise da produção de valor para vídeos digitais de música através da lógica social de derivativo. *Matrizes*, v. 12, n. 2, p. 193–215. doi:10.11606/issn.1982-8160.v12i2p%25p

van Dijck, J. (2014). Datafication, Dataism and Dataveillance: Big Data between Scientific Paradigm and Ideology. *Surveillance & Society*. doi:10.24908/ss.v12i2.4776

van Dijck, J. Van, Poell, T., & Waal, M. de. (2018). *The Platform Society: Public Values in a Connective World*. Oxford: Oxford University Press.

Drew, R. (2005). Mixed Blessings: The Commercial Mix and the Future of Music Aggregation. *Popular Music and Society*, 28(4), 533–551. doi:10.1080/03007760500159088

Drew, R. (2014). New Technologies and the Business of Music: Lessons from the 1980s Home Taping Hearings. *Popular Music and Society*, 37(3), 253–272. doi:10.1080/03007766.2013.764613

Frith, S. (1998). *Performing Rites: On the Value of Popular Music*. Cambridge, US: Harvard University Press.

Frith, S., Straw, W., & Street, J. (Eds.). (2001). *The Cambridge Companion to Pop and Rock*. Cambridge: Cambridge University Press.

Gurgel, D. (2016). O Novo Público Da Indústria Musical: Aquele Que Compra Ou Aquele Que Escuta? *Signos Do Consumo*, 8(2), 44–53. doi:10.11606/issn.1984-5057.v8i2p44-53

Gurgel, D. (2018). *A Imagem do músico em co-autoria com seu público: Uma análise da produção amadora de imagens através da ótica de sete artistas independentes e seus fãs*. (Masters thesis) – Escola de Comunicações e Artes, Universidade de São Paulo, São Paulo, BR. doi:10.11606/D.27.2018.tde-12072018-163840.

Hesmondhalgh, D., & Baker, S. (2011). *Creative Labour: Media Work in Three Cultural Industries*. New York, US: Routledge.

Hirsch, P. (1972, January). Processing Fads and Fashions: An Organization-Set Analysis of Cultural Industry System. *American Journal of Sociology*, 77(4), 639–659.

Hosokawa, S. (1984). The Walkman Effect. *Popular Music*, 4, 165–180. doi:10.1017/S0261143000006218

Ingham, T. (2018, February). The Great Big Spotify Scam: Did a Bulgarian Playlister Swindle their Way to a Fortune on Streaming Service? *Music Business Worldwide*. Retrieved from www.musicbusinessworldwide.com/great-big-spotify-scam-bulgarian-playlister-swindle-way-fortune-streaming-service/

International Federation of the Phonographic Industry – IFPI (2019). *IFPI Global Music Report 2019: State of the industry*. Zurich, CH: IFPI.

Jenkins, H. (2009). O que aconteceu antes do YouTube? In J. Burgess, & J. Green (Eds.). *YouTube e a revolução digital*. São Paulo, BR: Aleph, 143–164.

Katz, M. (2010). *Capturing Sound: How Technology Has Changed Music*. Berkeley: University of California Press.

Lewin, K. Psychological ecology. In: Gieseking, J. et al. (Eds.). *The People, Place, and Space Reader*. London: Routledge, 2014. p. 17–21.

Leyshon, A. (2014). *Reformatted: Code, Networks, and the Transformation of the Music Industry*. Oxford: Oxford University Press.

MBW. (2014). 'Taylor Swift is Absolutely Right': Spotify Flies Past 50M Users Worldwide. *Music Business Worldwide*. Retrieved from www.musicbusinessworldwide.com/spotify-flies-past-50m-users-worldwide/

McCourt, T., & Rothenbuhler, E. (1997). SoundScan and the Consolidation of Control in the Popular Music Industry. *Media Culture & Society*, 19, 201–218. Retrieved from http://hjb.sagepub.com.proxy.lib.umich.edu/content/9/2/183.full.pdf+html

McLuhan, M. (2001). *Understanding Media*. New York: Routledge Classics.

Morozov, E. (2013). *To Save Everything Click Here: The Folly of Technological Solutionism*. New York: PublicAffairs.

Morris, J. W. (2015). Anti-Market Research: Piracy, New Media Metrics, and Commodity Communities. *Popular Communication*, 13(1), 32–44. doi:10.1080/15405702.2014.977998

Morris, J. W., & Powers, D. (2015). Control, Curation and Musical Experience in Streaming Music Services. *Creative Industries Journal*, 8(2), 106–122. doi:10.1080/17510694.2015.1090222

Napoli, P. M. (2003). *Audience Economics: Media Institutions and the Audience Marketplace*. New York: Columbia University Press.

Negus, K. (1996). *Popular Music in Theory*. Hanover, US: Wesleyan University Press.

Negus, K. (2019). From Creator to Data: The Post-Record Music Industry and the Digital Conglomerates. *Media, Culture and Society*, 41(3), 367–384. doi:10.1177/0163443718799395

Nieborg, D.B., & Poell, T. (2018). The Platformization of Cultural Production: Theorizing the Contingent Cultural Commodity. *New Media and Society*, 20(11), 4275–4292. doi:10.1177/1461444818769694

Pariser, E. (2011). *The Filter Bubble. What the Internet Is Hiding From You*. New York: The Penguin Press.

Pettitt, T. (2012). Bracketing the Gutenberg Parenthesis. *Explorations in Media Ecology*, 11(2). doi:10.1386/eme.11.2.95_1

Poell, T., Nieborg, D., Duffy, B. E., Prey, R., & Cunningham, S. (2017). The Platformization of Cultural Production. *Selected Papers of #AoIR2017: The 18th Annual Conference of the Association of Internet Researchers*, (October), 1–18. Retrieved from http://platformization.net/wp-content/uploads/2018/09/AoIR2017-Platformization-of-Cultural-Production.pdf

Postman, N. (1993). *Technopoly: The Surrender of Culture to Technology*. New York: Vintage Books, pp. 71–72.

Prey, R. (2016). Musica Analytica: The Datafication of Listening. In R. Nowak & A. Whelan (Eds.). *Networked Music Cultures* (pp. 31–48). UK: Palgrave Macmillan. https://www.academia.edu/28851673/Musica_Analytica_The_Datafication_of_Listening.

Prey, R. (2018). Nothing Personal: Algorithmic Individuation on Music Streaming Platforms. *Media, Culture and Society*, 40(7), 1086–1100. doi:10.1177/0163443717745147

ProMusica (2019). *Mercado fonográfico mundial e brasileiro em 2018*. Rio de Janeiro, BR: ProMusica.

Radfahrer, L. (2018). O meio é a mediação: uma visão pós-fenomenológica da mediação datacrática. MATRIZes, v. 12, nº 1, p. 131–153. doi:10.11606/issn.1982-8160.v12i1p131-153

Sá, S, & Miranda, G. (2011). Aspectos da economia musical popular no Brasil: o circuito do Funk Carioca. In: Herschmann, M. (Ed.). *Nas bordas e fora do mainstream musical. Novas tendências da música independente no início do século XXI*. São Paulo, BR: Estação das letras e cores editora, 289–309.

Sauerberg, L. (2009). The Encyclopedia and the Gutenberg Parenthesis. *Media in Transition 6: Stone and Papyrus, Storage and Transmission*, April 24–26, 2009, Massachusetts Institute of Technology in Cambridge, MA, USA.

Seaver, N. (2017). Algorithms as Culture: Some tactics for the Ethnography of Algorithmic Systems. *Big Data and Society*, 4(2), 1–12. doi:10.1177/2053951717738104

Shoemaker, P., & Vos, T. (2009). *Gatekeeping Theory*. New York: Routledge.

Singer, J. B. (2014). User-Generated Visibility: Secondary Gatekeeping in a Shared Media Space. *New Media & Society*, 16(1), 55–73. doi:10.1177/1461444813477833

Vicente, E. (2012). Indústria da Música ou Indústria Do Disco? A questão dos suportes e de sua desmaterialização no meio musical. *Rumores*, 12, 1–16.

Vonderau, P. (2019). The Spotify Effect: Digital Distribution and Financial Growth. *Television and New Media*, 20(1), 3–19. doi:10.1177/1527476417741200

Wallace, J. (2017). Modelling Contemporary Gatekeeping. *Digital Journalism*, 6(3), 274–293. doi:10.1080/21670811.2017.1343648

White, D. M. (1950). The "Gate Keeper": A Case Study in the Selection of News. *Journalism Quarterly*, 27(4), 382–394.

Wikström, P. (2009). *The Music Industry: Music in the Cloud*. Cambridge: Polity Press.

Organizing music, organizing gender: algorithmic culture and Spotify recommendations

Ann Werner

ABSTRACT
Spotify is self-reporting to have 232 million monthly active users in July 2019, including 108 million paying subscribers. Often naturalized by listeners as a mere window into great collections of music, Spotify is an intricate network of music recommendations governed by algorithms, displayed as a visual interface of photos, text, clickable links, and graphics. With the aim to analyze how three Spotify functions, *related artists, discover*, and *browse*, organize and represent gender while organizing and representing music Spotify is here investigated through empirical material collected in qualitative online ethnographic studies during 2013–2015. The article problematizes how music is organized in algorithmic culture and uncovers gendering that can ensue as a result of the service's recommendation algorithms: creating closer circles for music consumption, and organizing music by similarities in genre and gender.

Finding new music and listening to music is a mediated process, whether we hear music on our stereo at home, in the supermarket, watch it on YouTube or in a live performance. Since media and media technology is indispensable in music listening it is constantly co-creating the experience of music (Taylor, 2001; Warner, 2003). While not always recognized as important for music listening, software is playing a central part of any computerized type of media output since web 2.0 (Chun, 2011). Scholars have labeled the marketing of cultural commodity like films, books, and music governed by software recommendations "algorithmic culture" (Galloway, 2006). Companies recommending books, like Amazon, films etcetera, like Netflix, or just any cultural commodity, like Google, are increasingly important in creating ideas about what cultural content is valuable and meaningful. Striphas (2015) has argued that value ascribed to culture is today determined by code hidden from us, owned by private companies. Music streaming services, like Spotify, are heavily relying on software algorithms protected by property rights to order and display its content. In this article, the software of Spotify is investigated through its effects, that is how content is organized to meet the listener by the functions *related artists, discover* and *browse*.[1] The Irish folk-rock artist Damien Rice serves as an example to enter the recommendation system of Spotify's algorithms.

Since software is integral for music listening today, one may wonder *how* software shapes the presentation of music in terms of power dimensions such as gender, sexuality, race/ethnicity, class, and nationality. Braidotti (2003, p. 61) argues that gender is a material experience *and* a symbolic construct. Gender organizes social and cultural life and is both felt and represented symbolically. According to her, the separation of the material from the symbolic and identity from society is false. Thus, feeling like a woman is not separable from ideas about femininity circulating at a certain time and place. Collins (1998, p. 63) has argued that gender is not distinct from other systems of oppression, like race and class. In fact, these systems articulate each other. When investigating the organization and representation of music as gendered it is therefore important to observe how gender intersects with for example race. Observing difference and same-ness is intrinsic to analyzing gendered culture. Braidotti (2003, p. 44) argues that ideas about difference within and between subjects are at the core of "gender" in culture and society. When difference is constituted between feminine and masculine so is similarity (between masculine and masculine). Gender as binary is central for meaning-making, according to Braidotti, even though binaries always are contested (Braidotti, 2003, p. 44).

Regarding material and symbolic dimensions of gender, feminist scholars in science and technology studies have studied how technologies are gendered. Technology intersects with the social and cultural formations of gender, in everyday use of technologies as well as processes of technological education and development of new technology (Wajcman, 2010, p. 144). Wajcman writes that in such processes "(n)either gender or technology are pre-existing" (Wajcman, 2010, p. 144). Here, the relationship between technology and gender as meaningful, and co-constituting is understood as involving multiple axes of power. Software, as a technology, is being co-constituted *with* gender, race, etcetera. Within a similar theoretical framework, McNeil (2007, p. 127) has argued that the masculinization of technology is an idea and practice not only evident in machines and institutions but also in symbols, language, and identity practices surrounding socio-technological relations. While McNeil, like Collins and Wajcman, does not see gender as an isolated power dimension she argues that the way technology is repeatedly masculinized is shaping technology as strongly gendered.

More specifically working on streaming software Eriksson and Johansson (2017, p. 177) has concluded that Spotify recommendations are dominated by artists labeled male in all genres and hereby furthers male-as-norm in the music industry. Their study shows that software is shaping socio-technological relations of gender for their users, making music streaming services worth discussing as technologies of difference in music today.

The aim of this article is to analyze how three Spotify functions, *related artists, discover*, and *browse*, organize and represent gender while organizing and representing music. Thus, how the software technology of the service genders the interface of music consumption. When the material for this article was collected *related artists, discover*, and *browse* were three prominent functions (Johansson, Werner, Åker, & Goldenzwaig, 2017). Drawing on a material of screenshots of particular pages and artists, and field notes taken while listening to Spotify on a daily basis, the article discusses how Spotify genders music. And how it builds on ideas of difference already existing in genre and artist representations of pre-streaming popular music culture.

These aspects are studied along three lines of inquiry: First the methodology, larger study, and context are presented. Then, critical analysis of music streaming and algorithms as well as current feminist studies of online media software are discussed as two research contexts shaping the analysis at hand. Next, the analysis is divided into three sections drawing on three Spotify functions, each deepening the understanding of how gendered software recommendations structure the interface on Spotify. And finally, two concluding sections focus on gendered music streaming as part of algorithmic culture.

Material and method

The material analyzed here is part of a larger study aiming to understand music use online. The material consisted of focus groups with 80 young adults in Moscow and Stockholm, and material from their favorite platforms. The platforms studied were Spotify, YouTube, and VKontakte. The focus groups were conducted in 2012 and 2013, while material was collected online between February 2013 and March 2015 in order to map the networks and the possibilities provided for online music (Johansson et al., 2017). The contexts of this study were two cities and digitalization of music culture. The study found that streaming has become an increasingly popular way to listen to, and discover music, especially in Stockholm but also in Moscow. In Sweden, Spotify was found to dominate music listening among the participants of the focus groups, while in Moscow VKontakte was the most popular site for music listening.

In this article, analysis begins by focusing on how Spotify software organizes a particular artist's network. This artist, Damien Rice, was discussed in some Stockholm focus groups as a good artist. Since Spotify adapts to the user by storing previous choices and basing suggestions on them it is hard to know what parts of the service are core-features, presented to all users, and what parts are individually adapted or location specific. To an unknown extent, recommendations discussed here have been adapted for me, based on my previous listening habits on Spotify. I joined Spotify as a paying subscriber in 2012 for research purposes but had been a nonpaying listener since 2009. Besides entry values such as where I live and my gender, what friends I had (Spotify was connected to Facebook in 2012), Spotify already had data about my listening habits. Research with bots (Eriksson & Johansson, 2017) has shown that Spotify recommendations tend to take on a gendered pattern for the listener independent of the given gender of the account holder but that genre listening shapes recommendations. Spotify is personalized and the organization of recommendations and relations between artists are based on accumulated choices of other users, thus, trying to find a "true" Spotify structure is impossible.

Mapping the logics of web sites through screenshots has been discussed as fruitful for digital methods by Moore (2014, p. 142) who argues that screenshots can be important analytical entities. Screenshots would not have been possible to analyze without the context of continuous use of Spotify over a period of time. The researchers' experiences were documented in field notes, common in online ethnography (Hine, 2015, p. 89; Werner, 2015). This article relies on these two groups of materials together with findings in the focus groups. The analysis is situated in contemporary studies of algorithms and software.

Algorithmic culture

Investigating software for music streaming Morris (2015a, p. 178) has argued that the business model of music streaming services relies on libraries of music being nontransferable to other platforms, effectively making one's record collection owned and organized by a commercial company. This, he claims, presents a game-changer for music culture. Software applications for music streaming change the ownership and durability of music listening, now highly vulnerable to the companies' policy development and commercial success. Further, he argues, recommendation systems give companies the power to determine the value of music as the new intermediaries of music (Morris, 2015b), gatekeepers of taste. Streaming services are central players in algorithmic culture, where cultural value is determined by processes partly human, partly machine. Mackenzie (2017, p. 8) argues that critical investigation of what commercial algorithms do has become a new strand of cultural theory devoted to machine learners. In an era where software is learning from experience, constructing complex systems functioning like black boxes (Mackenzie, 2017; Parisi, 2013) the logic behind recommendations on Spotify is difficult to fully know. Digital media is providing cultural content for streaming – for example watching and listening – but discoverability of content is heavily reliant on recommendations and search engines. These organize the large amount of content on Amazon, Netfix, and Spotify. Algorithmic forms of ordering cultural content has been labeled algorithmic culture, when algorithms decide ordering, visibility, and value ascribed to cultural commodity (Galloway, 2006; Hallinan & Striphas, 2016). Striphas (2015, p. 407) concludes that algorithmic culture presents itself as the effect of democratic processes, as if it is promoting culture consumed by many. While it is really private forces – companies – that are deciding the value of culture through software processes that remain hidden to us. Thus, algorithmic culture reduces decisions on taste to a few actors defining what is good and for whom, constructing social groups and cultural value in the process.

In studies of algorithms, it is relevant to ask: what combinations of software, hardware, and actors are encouraged and what does it mean? Someone accessing a website or service from a Swedish IP address is automatically assumed to understand Swedish and be interested in things Swedish, enforcing ideas on Swedish-ness in the process. Big data is not meaningful without the uses of it, as Brayne (2017) shows in her analysis of how police use big data in surveillance. She argues that the preexisting ideas that form what data is collected, for example who is considered a "risk individual" (Brayne, 2017, p. 986), are integrated into the effects of the data: who is surveilled and who is not. Assumptions about risk or more mundane things like national identity, language, and taste are not the only ideas algorithms speak to (and build on): through data mining users are mapped and approached accordingly in content as well as advertisement. The role of gendering in these processes has been studied but deserves more attention.

Feminist studies of software and social media

Linking software to power dimensions, Chun (2011, p. 2) has argued that software is omnipresent in new media but also impossible to map since software processes are infinite, and difficult to conceive, because of their invisibility to the user. Softwares are understood by Chun as mediums of power since they provide *one* way to navigate a complex machine world. The navigation software provides is not self-evident or neutral but based on choices.

One way that software helps the user navigate is by ordering identities in systems of gender and race. Nakamura (2002) argues that online gendered and raced identities often are presented as "menus" to choose from. At the time of her research, such choices were often visible in the interface in games or chatrooms and after the user made their choices the software used them to present the user with some options and not others. Nakamura concludes that identities are perceived as clear-cut choices in this type of software, man/

woman, white/black/Asian, making it impossible to represent the fluidity of everyday identity, identifying for example as both Asian and black, or neither man or woman. Identity consisting of one gender, one race, one hometown, is what the software built on to understand users. Software's way of coding identity selections was integral in the shaping of web 2.0 (Nakamura, 2002). These choices have stayed on in the development of social media even though they may be less visible. McNicol (2013, p. 205) states that gendered choices are inscribed differently in social media systems, choices are more or less required by different services and the options are formulated differently on social media like Facebook, Twitter, and Diaspora. Therefore, the effects of gendering on social media can also be different. Bivens (2015) has investigated Facebook software paying interest to how Facebook is organized as gender binary. She has argued that one effect of the algorithm *people you may know* on Facebook is that it puts survivors of sexual harassment and violence in contact with their perpetrators, possibly prolonging the trauma. Paying attention to gendering effects of software in social media Bivens (2015, p. 715) argues that social media software is ripe for feminist analysis, and that social media software is deeper manifestations of power than the surface content. In her research on Facebook Bivens (2015, 2017) investigates how software enacts normalizing logics, for example by offering several choices for gender identity while still hiding a binary division of users in terms of gender in the original code.

The secrecy may also be analyzed as part of the meaning of software (Chun, 2011, p. 5). The interface is presented as neutral, normal, when in fact it is the result of a series of choices made with purposes that are hidden and assuming difference in terms of identity. Bishop (2018, p. 81) has argued that the algorithm of YouTube rewards normative white femininity in beauty vloggers by recommending it more often, in line with desires of brands and advertisers who deem this femininity as more marketable. It does so by the choices made by the algorithms, notably what videos are suggested to viewers watching beauty vloggers.

Three functions

On Spotify, music listeners can make playlists, save favorite artists, and albums in their archive called *my music* and listen to radio channels, or playlists put together by themselves, other listeners, or by Spotify. In order to discuss how recommendations are structured as socio-technological, and as such building on and furthering some ideas of gender intersecting with race and nationality, I will address three functions on Spotify. The functions are: *related artists* that lists artists understood as similar to the one you are already listening to, *discover* that suggests personalized music based on previous listening and what Spotify deems similar to that listening, and the first page called *browse* that assembles Spotify's most important features.[2] This selective method of approaching Spotify through functions was motivated by the central place of these functions at the time of the study and how they were discussed in focus groups.

Who is kin? The *related artists* function

On Spotify, at the time of the study, the *related artists* function suggested artists presented as similar to the one you were currently listening to. While all three functions analyzed in this article were modified during 2013–2015 the main principle of them remained intact.

In some ways, they stay important for Spotify in 2019 even though they are labeled differently (*related artists* is now called *fans also like*) and their placement in the interface differs (*browse* has been moved, *discover* is now *discover weekly*). The idea that music is connected to other music is the core of Spotify's ordering of music through recommendations in *related artists* and *discover*. In the first year of material collection, I listened to Irish folk-rock singer Damien Rice – because he was popular among the participants in the Stockholm focus groups. When I first opened Damien Rice's profile I noticed that all the artists presented by *related artists* were white-ish men – not a noteworthy or unusual observation in itself. Feminist music scholars have concluded that the genre rock is masculinized, white washed, and also dominant on the western popular music market and in music journalism (Coates, 1998; Leonard, 2007). The most popular rock artists are often presented as white men, solo artists or in groups, even though the type of masculinity they feature varies in different types of rock, from metal to folk. Artists identified as women, often but not always representing femininity, are also present, but often marginal in rock. Music genres interpreted as masculine, like rock, have often been considered of higher value and more authentic than feminized popular music genres like pop (Biddle & Jarman-Ivens, 2007, p. 3).

The idea that some artists were related intrigued me, the word "related" implies kinship, kinship in turn is often imagined in terms of gender, nation, and race. And while white rock music seldom is described as white, it can be labeled as national, like for example British rock and pop music (Stratton, 2010, p. 27, see also Zubeeri, 2001). Stratton (2010) argues that when calling music "British" there is an underlying assumption that British means white – if a group or artist is not white this has to be pointed out. Ascribing music to a certan nation also usually takes the sounds of that music into account. Music may be percieved as sounding British, like brit pop, or not, like bhangra, both genres are produced in the UK but understood differently in terms of nation and race. Artists that are seen as related by Spotify may therefore be so – in terms of sharing group identities based on gender, nation, class, and race. The white folk-rock men connected to Damien Rice by Spotify caught my interest and I started keeping track of the network of artists unfolding with Damien Rice as starting point. He will be used as an example to discuss the *related artists* function.

I concluded that Damien Rice was presented as male by Spotify because he was referred to as a he, and so were most artists suggested by *related artists* on his page. Artists not referred to as he or she are uncommon on Spotify, still, in cases such as popular transgender artists like Antony Hegarty (of Antony and the Johnsons) Spotify refers to Antony by name, not using any pronoun in the text describing them. Spotify operates with a binary idea of gender while recognizing some openings for alternatives outside the binary, without naming them. Race was not explicitly addressed in the descriptions of the majority of artists related to Damien Rice. And since they appeared white in their photos not addressing whiteness reinforces white as an invisible norm, reinforcing the idea that white people are "outside" of race, and race belongs to the other. Sometimes nationality was used to describe artists, "Irish", or "Czech" (while Tracy Chapman was "African-American", a racial signifier), and their nationality could, as Stratton (2010) argues, be seen to contain a racial presentation, Irish implies white and Irish unless some other feature is presented. Sometimes class was mentioned in terms of background, when artists were described as having a "working class background".

In the first months of 2013, I followed Damien Rice's *related artists* around by mapping their (in turn) relations in three steps and observing genre affiliations as well as representations of gender, race, and nationality in their pictures, names, and biographies. I also listened to their music, even though this article does not focus on the in-music qualities of sounds and lyrics. My mapping showed that predominantly male artists, predominantly white artists, predominantly singer-songwriter and folk-rock artists and predominantly solo artists were related to Damien Rice within three steps of recommendations.[3] Musically a slow tempo folk-rock featuring acoustic elements and emotional lyrics (often about romantic love and loss) in the English language dominated. The style of most popular songs of these artists contributed to presenting a brooding, emotional, and sensitive yet strong masculinity easily recognizable from singer-songwriter and folk-rock genre tradition. The nationalities of the artists related to Damien Rice were mainly European or North American. The age and generation of the artists varied: older artists not actively writing music anymore like Bob Dylan and Nick Drake appeared, so did Glen Hansard, The Frames and David Gray and even though there was an Irish white dominance a Swedish artist with parents from Argentina, José Gonzales, and a female African-American artist, Tracy Chapman could both be found in the network (they were presented in this way). The artists considered related to Damien Rice by Spotify were partly changing over time, while the overall impression of genre, nationality, race, and gender in the group stayed the same, and some artists, like Glen Hansard, were always in the first step after Damien Rice.

Later on in the study, in March 2015, there was one female white solo artist in the top four *related artists* to Damien Rice: Lisa Hannigan who is Damien Rice's ex-girlfriend. In February 2013, there was one artist presented as female in the top five *related artists* to Damien Rice: Marketa Irglova (one half of The Swell Season, a group also within the network). Marketa Irglova performed alongside Glen Hansard in the popular Irish movie *Once*: portraying love and Irish folk-rock music. Apart from these two women, Tracy Chapman, and some Irish folk-rock bands with male and female members, very few female artists appeared within three steps of Damien Rice during the years of study, even fewer black or brown artists appeared (only the two artists already mentioned). The *related artists* function is according to an interview with Spotify founder Daniel Ek (Gelin, 2012) based on accumulation of choices that is fed back to listeners. This is a common practice in algorithmic culture, and his explanation does not give away what choices the algorithms are making. The software appears neutral, and is unknown, but based on multiple choices fed into preexisting number of programming presumptions. As Striphas (2015) has argued the recommendations do not only mirror accumulation of choices, it builds on the companies definitions of commodities, social groups, and categorizations of both, genre being important on Spotify.

In the focus group, study participants described the *related artists* function as a good way to discover new music. Finding new good things to listen to is every music lover's dream: therefore, the *related artists* function as such shapes the use of Spotify. In the case of Damien Rice, the function promoted choices that were similar to Damien Rice in terms of how genre, gender, and race were presented. The latter two categories take shape through the genre category. Genre rules associate the style of folk-rock Damien Rice performs with Irish, British, and North American nationality, whiteness, and masculinity. But there are many more female (mostly white) folk-rock singer songwriters that do not appear around Damien Rice, thus genre as explanation does not fully hold up. As Striphas (2015) has argued algorithmic culture hides an elite system of unknown choices made by algorithms behind the presumption that the "users" are in charge of culture through the accumulation of choices. The presentation of recommendations and top choices in cultural consumption presents ideas about cultural value, as well as of social groups (Striphas, 2015, p. 406). In *related artists*, cultural value around Damien Rice is white and male, with few exceptions. And the female artists often suggested by *related artists* (Lisa Hannigan and Marketa Irglova) are also presented as romantic partners to Damien Rice in real life and Glen Hansard on screen, rather than successful on their own terms.

Who are you? The *discover* function

The *discover* function presented a list of personalized suggestions, you could at the time of study find it on the *browse* page, click on it and enter personalized suggestions. It recommended artists and songs to listen to, based on previous account activity, the collected habits of all listeners and constructed ideas about music similar to the one you liked (thus seemed to build on *related artists*). As *discover* is personalized – adapted to previous patterns on the account – and *related artists* seem to be based on national user habits (Gelin, 2012), my use of Spotify at the time was providing a unique interface. In short, based on my listening habits, nobody else would get the exact same suggestion pattern as I did. On the other hand, the pattern displays the network of artists and genres that Spotify recommendations build on and some of the re-occurring principles of the algorithms. Thus, while geographical points of reference and my already existing choices are particular the logics of *discover* display algorithms all listeners meet.

Discover implies, by name alone, that there is a possibility for the listener to discover new music. But the suggestions made to me during the period of the study were rarely artists that were unknown to me. Suggestions like "You listened to Beyoncé maybe you would like Brandy" can almost seem to mock the listener, and *discover* rarely led me to discover artists I did not know. In order to experiment with the *discover* function, I listed to Spotify every day, according to my own taste in periods and to particular artists and genres I would not normally listen to in other periods. During the autumn of 2014, I listened to a large amount of K-pop, mostly groups with several female members. Within a day *discover* suggested young South Korean female pop artists back to me, and a few young men. The artists Spotify recommended to me were popular chart successes in South Korea, but not all of the songs were recent releases, this can be understood as a result of my national position, I might have gotten more up-to-date suggestions had I been on a South-Korean IP address since national listening patterns are used in the recommendation algorithms. Accordingly, when I stopped listening to K-pop the suggestions in *discover* changed. Thus, *discover* was in 2014 a very temporally sensitive function picking up on recent listening habits. Earlier on in 2013 *discover* would sometimes suggest music that I had not listened to in a long time, also telling me so "you have not listened to x in a while". This highlights that Spotify had been tuning the algorithm during the period: experimenting to find the now-ness of recommendations. Being able to use surveillance knowledge in real-time is a consequence of big data (Brayne, 2017, p. 991). It is easy to monitor a certain period, and the present is often seen as more relevant than the past. While *related artists* was a function producing patterns that did not change that much over time (around Damien Rice) *discover* was time sensitive and picked up on recent habits of the individual listener. While this distinguished *discover* from *related artists* similarity in genre, gender, and race was recommended by both functions.

We are Spotify. The *browse* function

The first page of Spotify was called *browse* and contained a banner with announcements on top, and then content consisting of an overview of popular *playlists, new releases, news, top lists, genres, moods,* and the *discover* function. *Browse* promoted the main guiding functions, highlighting some *playlists, new releases,* and *genres.* Opposed to early YouTube's centrality of the *search* function, Spotify aimed to help the listener by arranging the music like in a record store (but Spotify has a *search* function too). Fore fronting some genres and new releases remediated the organization of a physical record store. The *playlist* was inspired by early iTunes, and the *discover* function can be seen as being based on Pandora's business idea created in 2000, aiming to find the next similar (but not too similar) song. Avdeeff (2012) concludes that personalized music consumption has grown with digital music use. Her research shows the ease with which a listener can personalize one's listening, picking the next artist or song on the go, the listener understands this as an advantage of digital music use: having a record collection to choose from in one's pocket. *Browse* was a function that aimed to present and organize this collection of music for the listener.

Browse was clearly shaped by the day and time one entered it: by advertising playlists like "Morning commute" in the morning and "New music Friday" on Fridays, and by the news (Avicii releases a new single on so-and-so date). The top advertisement banner took up a lot of space in *browse* and often presented a new album or playlist. All of these ways of guiding the user can be read as aiming to create an experience of time. *Browse* presented Spotify as a place within time. It strived to be part of the listeners' everyday rhythm and up-to-date (see Johansson et al., 2017, chapter 2). While *browse* differed between countries, by presenting popularity of songs divided by countries under *top lists,* it seemed to appear the same for all users in one country (except in *discover*). *Browse* did not order suggestions by similarity. It presented a variety of artists and genres, while at the same time still portraying ideas of genre, gender, and race. For example: the genre soul was during the study represented by a logo shaped as a woman's head – with afro hairstyle and hoop earrings implying a black woman – while rock was represented by an amplifier. Music genres are established by rules that are based on social understandings (Fabbri, 1981), and as such not given by musical sound or style only. Still, stylistic components in the music do play a part in genre classification. Genre groupings also often hinge on for example nation, race, class, gender, and sexuality (Brackett, 2016, pp. 3–4). To illustrate playlists found under the link to the genres soul and rock photos were used. The top playlists under the genre soul were illustrated with African-American artists, often men and women smiling, while the top playlists under the genre rock were illustrated with guitars, empty houses, and white men. Ordering and representation were significant for how Spotify portrayed music for the listener entering through *browse,* even though *browse* displayed different routes available. The routes the listener could take were often illustrated visually through ideas about different social groups representing different music genres. Further, most choices Spotify's software made when targeting the Swedish audience on *browse* through news promoted Swedish artists and artists from the English-speaking world. National diversity was not promoted by *browse.* There was a mainstream dominance promoting pop, rock, and EDM but the artists recommended were not necessarily only mega stars of chart pop. The promotions can be assumed to mirror expectations of the Swedish Spotify listeners' habits by A & R and record companies. Without being sure about how the software selects, and in what pool of different choices the selections are made, *browse* built an idea of Swedish taste and created patterns of listening, by urging listeners to notice for example Avicii's new single. Both the invisible algorithmic choices, and the visible interface design and its portrayal of music and artists nudged the listener in different directions according to taste and identity. Still, the logic of *browse* was more diversified than the two functions previously discussed and it seemed that *browse* aimed to speak to multiple (Swedish) audiences at the same time.

Gendered music streaming

The functions *related artists, discover*, and *browse*, as has been shown, strengthen ideas about genre affiliations of artists and songs. They render presumptions about taste in your nation of IP origin important but invisible. They do this by feeding back other listeners' choices, algorithmic choices and choices made on your account as recommendations of new music. These recommendations are ordered in terms of similarity, this similarity is musical but also include similar representations of gender and race. What possible music listening does this create? One answer is that the functions result in a feedback loop: following *related artists* from Damien Rice led me back to Damien Rice every step of the way. The use of genre and previous choices as cornerstones in Spotify's construction of similarity are here found to reinforce connections between artists similar in terms of gender and race. The recommendation functions on Spotify make similarity central and in that process they emphasize difference. When Tracy Chapman appears as related to Damien Rice she really stands out as a black woman. While genres can be broad, *related artists* creates networks where most artists are very similar to the first artist of choice. For example: Rihanna leads you to the top-related artist Beyoncé, who leads you to Destiny's child, bringing you to Kelly Rowland who leads to Ciara and Ciara leads you back to Kelly Rowland. No male artist, white artist, and no artist from another generation, nation, or genre are included in this circle of recommendations.

The recommendations of Spotify reflect patterns already known, premieres very famous artists and rarely gives surprising advice. Also, since the recommendation system is accumulated over time few debuting artists will appear though *related artists* or *discover*, and neither will old forgotten artists.[4] Rose (1994) has argued, in relation to hip hop, that musical genres can be racialized in a politically progressive way, fighting racism, and simultaneously contain sexist messages fixing gendered racial stereotypes. Hence, gendering and racialization of genres may have progressive potential and essentializing dangers at the same time. What can be observed in Spotify's functions is that genres become narrowly gendered through *related artists* and *discover*. In a record store Beyoncé or Ciara might be placed next to a male artist within mainstream R&B, or a not so popular artist – they would still likely be categorized as part of an African-American popular music tradition – but on Spotify *related artists* lists them next to young African-American top-selling females. Participants in focus groups argued that they could find immense amounts of music through the Internet, and researchers have pointed out a heightened possibility for musical eclecticism supported by digital music formats (Avdeeff, 2012). But, it seems doubtful that the software of contemporary streaming services supports these practices. While listeners may perceive Spotify as liberating, and use it in such ways, the functions analyzed here limit listening patterns by building gendered and racialized connections between artists and genres. Users of Spotify may still listen to a variety of music that they find out about for example through other media, through advice from friends or by going to live performances. Also, they may actively use the search function to find artists or explore different genres starting through *browse*. But this is not what Spotify invites them to do.

Conclusion

The business idea of companies selling streamed music has been scrutinized for its capitalist logics (Morris, 2015a, 2015b). Claims are that algorithmic culture hides how companies shape cultural value and social groups through the algorithms while presenting them as "choices" of audiences (Striphas, 2015). By promoting similarity, emphasizing what is popular right now and building on the genre system, Spotify's network of music aims to simplify listening. At the same time, Spotify organizes gender, nationality, and race in music culture. This is done materially by connections in the interface, and discursively by representations of artists and genres. While harmless on the surface, when considered more closely Spotify's functions for recommendations help reconstruct dominating genres like rock as male-focused, masculine, and white. This gendering of rock has material implications in listening experiences and further a masculine rock discourse. While the aim of Spotify may be to personalize listening and guide their users to music they will love, the result is enhancing the already existing gendering of popular music genres. Acknowledging that popular music genres are shaped by nation, race, gender, etcetera (Brackett, 2016), they also articulate historical injustices such as colonialism, slavery, sexism and oppression of women, and violence against transgender persons. The organizing and representation of musical taste and social groups on Spotify are thus not coincidental, or innocent, but reinforcing patterns of power.

Notes

1. Empirical material discussed here was collected within the research project "Music use in the online media age," 2012–2015, funded by Riksbankens Jubileumsfond, and conducted with Sofia Johansson, Patrik Åker and Gregory Goldenzwaig. The project conducted focus groups and online ethnography (on VKontakte, Spotify, and YouTube) mapping the meaning of music online. The main results are discussed in *Streaming Music* (Johansson et al., 2017).
2. The functions described here mirrors Spotify in 2014–2015 as it looks in a computer interface for paying subscribers. The material about the *related artists* function was collected over a longer period of time, starting in February 2013.
3. Male artists dominate popular music (Smith et al., 2019) and most studies put women/others at around 20% among performing artists. The argument here is not that algorithmic culture distorts reality (or not). How algorithmic culture is gendered and what is presented as recommended music is in focus.
4. Debuting artists may appear on *browse*.

Disclosure statement

No potential conflict of interest was reported by the author.

References

Avdeeff, M. (2012). Technological engagement and musical eclecticism: An examination of contemporary listening practices. *Participations: Journal of Audience and Reception Studies, 9*(2), 265–285.

Biddle, I., & Jarman-Ivens, F. (2007). Introduction. Oh boy! Making masculinity in popular music. In F. Jarman-Ivens (Ed.), *Oh boy! Masculinities and popular music* (pp. 1–20). New York: Routledge.

Bishop, S. (2018). Anxiety, panic and self-optimization: Inequalities and the YouTube algorithm. *Convergence, 24*(1), 69–84. doi:10.1177/1354856517736978

Bivens, R. (2015). Under the hood: The software in your feminist approach. *Feminist Media Studies, 15*(4), 714–717. doi:10.1080/14680777.2015.1053717

Bivens, R. (2017). The gender binary will not be deprogrammed: Ten years of coding gender on Facebook. *New Media & Society, 19*(6), 880–898. doi:10.1177/1461444815621527

Brackett, D. (2016). *Cathegorizing sound: Genre and twentieth-century popular music*. Oakland: University of California Press.

Braidotti, R. (2003). Becoming woman: Or sexual difference revisited. *Theory, Culture & Society, 20* (3), 43–64. doi:10.1177/02632764030203004

Brayne, S. (2017). Big data surveillance: The case of policing. *American Sociological Review, 82*(5), 1–32. doi:10.1177/0003122417725865

Chun, W. H. K. (2011). *Programmed visions: Software and memory*. Cambridge, MA: MIT Press.

Coates, N. (1998). Can't we just talk about music: Rock and gender on the Internet. In T. Swiss, J. Sloop, & A. Herman (Eds.), *Mapping the beat: Popular music and contemporary theory* (pp. 77–99). Malden, Mass: Blackwell.

Collins, P. H. (1998). It's all in the family: Intersections of gender, race and nation. *Hypatia, 13*(3), 62–82. doi:10.1111/j.1527-2001.1998.tb01370.x

Eriksson, M., & Johansson, A. (2017). Tracking gendered streams. *Culture Unbound, 9*(2), 163–183. doi:10.3384/cu.2000.1525.1792163

Fabbri, F. (1981). A theory of musical genre: Two applications. In D. Horn & P. Tagg (Eds.), *Popular Music Perspectives* (pp. 52–81). Göteborg and Exeter: International Association for the Study of Popular Music.

Galloway, A. R. (2006). *Gaming: Essays on algorithmic culture*. Minneapolis: University of Minnesota Press.

Gelin, M. (2012, December 2). Så bygger Daniel Ek musikens framtid. *Dagens Nyheter*, K12. (This is how Daniel Ek builds the future of music).

Hallinan, B., & Striphas, T. (2016). Recommended for you: The Netflix prize and the production of algorithmic culture. *New Media and Society, 18*(1), 117–137. doi:10.1177/1461444814538646

Hine, C. (2015). *Ethnography for the Internet: Embedded, embodied and everyday*. London: Bloomsbury Academic.

Johansson, S., Werner, A., Åker, P., & Goldenzwaig, G. (2017). *Streaming music: Practices, media, cultures*. Abingdon, Oxon: Routledge.

Leonard, M. (2007). *Gender in the music industry: Rock, discourse and girl power*. Aldershot: Ashgate.

Mackenzie, A. (2017). *Machine learners: Archeology of a data practice*. Boston, MA: MIT Press.

McNeil, M. (2007). *Feminist cultural studies of science and technology*. London: Routledge.

McNicol, A. (2013). None of your business? Analyzing the legitimacy and effects of gendering social spaces through system design. In G. Lovink (Ed.), *Unlike us: Social media monopolies and their alternatives* (pp. 200–219). Amsterdam: Institute of Network Cultures.

Moore, C. (2014). Screenshots as virtual photography: Cybernetics, remediation, and affect. In P. L. Arthur & K. Bode (Eds.), *Advancing digital humanities: Research, methods, theories* (pp. 141–160). Basingstoke: Palgrave Macmillan.

Morris, J. W. (2015a). *Selling digital music: Formatting culture*. Oakland: University of California Press.

Morris, J. W. (2015b). Curation by code: Infomediaries and the data mining of taste. *European Journal of Cultural Studies, 18*(4–5), 446–463. doi:10.1177/1367549415577387

Nakamura, L. (2002). *Cybertypes: Race, ethnicity and identity on the Internet.* New York, NY: Routledge.

Parisi, L. (2013). *Contagious architecture: Computation, aesthetics and space.* Cambridge, Mass: The MIT Press.

Rose, T. (1994). *Black noise: Rap music and black culture in contemporary America.* Hanover, N. H.: Wesleyan University Press.

Smith, S. L., Choueiti, M., Pieper, K., Clark, H., Case, A., & Villanueva, S. (2019). *Inclusion in the recording studio? Gender and race/ethnicity of artists, songwriters and producers across 700 popular songs from 2012-2018.* Los Angeles, California: USC Annenberg inclusion initiative.

Stratton, J. (2010). Skiffle, variety and Englishness. In A. Bennett & J. Stratton (Eds.), *Britpop and the English music tradition* (pp. 27–40). Farnham: Ashgate.

Striphas, T. (2015). Algorithmic culture. *European Journal of Cultural Studies, 18*(4–5), 395–412. doi:10.1177/1367549415577392

Taylor, T. D. (2001). *Strange sounds: Music, technology and culture.* New York, NY: Routledge.

Wajcman, J. (2010). Feminist theories of technology. *Cambridge Journal of Economics, (2010*(34), 143–152. doi:10.1093/cje/ben057

Warner, T. (2003). *Pop music, technology and creativity: Trevor Horn and the digital revolution.* Aldershot: Ashgate.

Werner, A. (2015). Moving forward: A feminist analysis of mobile music streaming. *Culture Unbound, 7*(2), 197–213. doi:10.3384/cu.2000.1525.1572197

Zubeeri, N. (2001). *Sounds English: Transnational popular music.* Urbana, Ill.: University of Illinois Press.

What do we do with these CDs? Transitional experiences from physical music media purchases to streaming service subscriptions

Waleed Rashidi

ABSTRACT

As streaming music service subscriptions quickly become a primary channel for listeners, how are previously popular channels—namely physical artifacts of music recordings—complemented or replaced? A survey with qualitative responses was conducted with streaming music subscribers to understand if physical media such as compact discs are still used, if music collections exist (if not, how they were discarded), and if subscribers plan to purchase physical artifacts of music in the future. Results indicated that most subscribers still owned physical recordings; unwanted physical recordings were usually donated or given away. A majority did not plan to purchase more physical recordings; however, expansion of collections, nostalgic engagement, and a preference for tangible formats were some future purchase reasons by current subscribers. The results reflected consistencies with earlier studies examining audiences of recorded music.

Introduction

The growth of streaming music services over the past several years has led to a revolution in which context consumers listen to pre-recorded music. The subscription model of music playback and free-to-stream music videos on YouTube may have hastened the diminution of the once-dominant physical format, the compact disc, which—prior to the proliferation of the iPod and downloadable MP3 file—remained popular in the music marketplace for some two decades (Knopper, 2018). Increasingly, audiences are adopting streaming services as conduits of music listening. The International Federation of the Phonographic Industry's (IFPI) Global Music Report 2019 indicates that global music subscription audio streaming's revenue share—at 37%—outpaced physical recording's revenue share at 24.7% in 2018; paid audio streaming revenue increased 32.9% over the previous year, while physical revenue declined 10.1% (IFPI, 2019).[1] For some, purchasing CDs or vinyl LPs from stores have been supplanted by a monthly bill and immediate streaming content from online services. This move from a culture of ownership of physical music artifacts to a culture of access via music streaming services is not exclusive to the music industry, as de-materialization can also be found in transportation (rideshare companies such as Uber or Lyft) and audiovisual playback (for example, video streaming via Netflix) (Given, 2015; Herbert, Lotz, & Marshall, 2019; Sinclair & Tinson,

2017). Given's (2015) historical examination of rationales for ownership versus access of audiovisual media noted that such change was interpreted by some as "...part of a longer-term, economy-wide and worldwide shift away from ownership towards access models" (p. 36).

A weekly advertisement on Target.com for Target's Black Friday sale in the U.S. in November 2018 featured several vinyl LPs of major artists including Bob Marley, Guns N' Roses, and Amy Winehouse, but no CDs were shown. Target was reported by *Billboard* to recently stock only about 100 CD titles, down from approximately 800 (Kumar, 2018; Mervis, 2018). Best Buy, another U.S. retailer whose stores once housed several CD aisles, recently featured just one row of CDs for sale (Kumar, 2018). Independent retailers noted that CD purchasers are typically older adults seeking blues, soul, and classic rock (Mervis, 2018).

As music audiences continue in this technological shift towards streamed playback, a question remains: What becomes of the physical formats—specifically CDs—that were once the sole means of music playback? This exploratory study examines what music audiences in the U.S. have done with their physical media, after subscribing to and making streaming audio services their preferred source for recorded music. Incorporating a uses and gratifications approach (Katz, Blumler, & Gurevitch, 1974), the study assesses what U.S.-based streaming music users are doing with their physical media collections and what users derive from such format engagement. This exploration of format preferences is consistent with other U&G media studies (examples are outlined in Ruggiero, 2000), including Lonsdale and North's (2011) study of the reasons people listened to music, and Hollebeek, Malthouse, and Block's (2016) research in music engagement.

The present investigation is guided by several prompts: Are listeners' CDs and vinyl LPs still in use? Why do certain titles remain in listeners' physical media collections? How do listeners decide which physical albums to keep and remove? And do these subscribers envision themselves purchasing physical albums in the future?

Literature review

Even studies of online music usage less than a decade old may hold less relevance today, as many could not predict the rapid popularity of music streaming services. Spotify, the largest of these platforms, began offering its service in Europe in 2008, with a U.S. entry in 2011 (Swanson, 2013). A streaming service analyst stated that 202 million subscribers had paid for music streaming services at the end of 2017 (Roberts, 2018). And a report notes that Spotify's subscriptions climbed 10% in mid-2018, with 83 million subscribers in total (Steele, 2018); Spotify had 100 million paid subscribers as of April 29, 2019 (Kelley, 2019), with a 36% share of global music streaming subscribers, followed by Apple Music (19%), Amazon (12%), and Tencent (8%) (Watson, 2019a). Apple Music has over 2 million more U.S.-based paid subscribers than Spotify (Palladino, 2019). Deezer has grown in the past few years, particularly in France and the Netherlands (Watson, 2019b). And YouTube Music (owned by Google), which was introduced in June 2018 with three tiered prices, offers ad-free content uploaded to the YouTube platform (Roberts, 2018). While YouTube itself is free (with embedded advertising) and is the most widely used video hosting website, especially music video content (Liikkanen & Salovaara, 2015), Google's paid music subscriber counts via services like YouTube Music trail considerably behind both Spotify and Apple Music (Shaw, 2019). Still, YouTube remains a formidable repository of both major label and independent artists' videos (Edmond, 2014). Tidal, a comparatively smaller service than Spotify or Apple Music, has taken a unique position of being a platform of exclusive content without a standard free user tier (Meier & Manzerolle, 2019); users can access the service on a free 30-day trial.

Swanson's (2013) study examined users' selection of streaming services, finding that listeners opted for convenience, quality, and decreased cost (64% used free services). YouTube could act as a promotional vehicle, directing users towards music sales via its recommendations (Hiller, 2016). Still, Swanson (2013) found that only 18% of the 18–24 age set reported that Spotify usage altered music purchase habits either "drastically" or "a lot" (p. 217).

Nevertheless, music purchase habits are changing. While three-quarters of the overall revenue generated from recorded music in the U.S. hail from streaming services, vinyl gained nearly 13% from the year prior (Recording Industry Association of America, 2018). Digital downloads in the U.S. accounted for 12% of revenues, and CD sales dropped over 41% (Recording Industry Association of America, 2018). The CD is quickly losing status as the preferred music playback option.

The CD player reached its Australian household diffusion peak in 1993 (Ironmonger, Lloyd-Smith & Soupourmas, 2000). Citing Recording Industry Association of America (RIAA) figures, CD sales have declined 87% in 15 years, from 942 million units in 2000 to 125 million by 2015 (Mitchell, Scott, & Brown, 2018). CD sales fell below 100 million for the first time in the current decade, to 93 million sold in 2016, according to a Nielsen Music report (cited in McIntyre, 2017). More recently, U.S. CD sales fell 6% in 2017 (Kumar, 2018). The planned demise of the CD by major record labels was reported several years ago (Miers, 2011; Thrasher, 2012), though the phase-out would likely be a long-term transition as new CDs are still available for sale.

One issue with CD sales is the value proposition, as some listeners may not feel they are worth the price. Consumers who were price conscious were more inclined to engage in music streaming or downloading for media delivery as opposed to the purchase of a CD (Im & Jung, 2016). And a study found that concert attendees were willing to pay over 50% more for a concert ticket than their last concert purchase, yet CD purchasers were only willing to spend about 5% more than their last CD purchase, "…perhaps influenced by the new cheaper technologies to download and play music" (Rondan-Cataluna & Martin-Ruiz, 2010, p. 1418). However, Sinclair and Tinson (2017) found that fans' financial and time investments in—and motivations for—establishing a physical record collection meant that some consumers still find value in ownership.

Another issue is that as technology progresses, the CD is becoming less commonplace; this is noticeable within automobile stereos, as pre-installed CD players are less frequently available as audio options for new cars. Major brands such as Ford, Honda, Mazda, Chevrolet, and Toyota ended factory installation of CD players in some new models; nearly half of all new cars sold in the U.S. by 2021 are predicted to not include CD players (Ford-Stewart, 2018; McIntyre, 2017). Given that music is often played in automobiles, this is an important point when considering the CD's longevity as a format.

However, for consumers who currently purchase physical artifacts of music, a market still exists. Reasons cited for current CD sales include enhanced sound quality and better affordability due to lower demand (Kumar, 2018). Sound quality may also influence CD usage over illegally downloaded music (Poort & Weda, 2015). And consumers who maintained music catalogs preferred to purchase CDs, as opposed to legally downloaded purchases or streams (Im & Jung, 2016).

The general decline of physical music sales did not deter Taylor Swift from promoting sales of her own CD. The popular musician first released *Reputation* in November 2017 on physical formats (and digitally purchased downloads), but delayed the album's availability on streaming services until the following month (Stutz, 2017). Swift released two physical versions of *Reputation* via Target, each featuring a different magazine, which encouraged some fans to make two CD purchases in order to own both magazines (Steele, 2017). Other artists have opted to forgo a CD release all together, pressing vinyl-only physical recordings, including a live album by Jack White (Sodomsky, 2018) and Jeff Tweedy's *Warmer* (Kreps, 2019). There are also entire record labels dedicated to vinyl-only releases (Sherburne, 2016).

Music rights holders (including record companies) have lamented the loss of physical sales, as they receive lower royalties from streamed plays (Thomes, 2013; Tilley, 2018). Still, peer-to-peer (P2P) file sharing networks—ubiquitous in the late 1990s and 2000s—assisted in the decline of physical format sales (Kot, 2003). Even giving customers the option to purchase tracks downloaded via online retailers was not enough to curb piracy's impact (Waldfogel, 2010). Although legal streaming services have been noted as displacements for music piracy portals (Aguiar & Waldfogel, 2018), Borja and Dieringer (2016) found that not only does streaming service use increase the possibility of piracy, but that college students surveyed did not see streaming as a cost-effective piracy replacement.

So, who still purchases physical artifacts of music? As of 2015, Japanese consumers—the second largest music market in the world after the U.S.—purchased over 80% of their music on physical formats such as CDs (Sisario, 2015). Dilmperi, King, and Dennis (2011) found that the older the listener, the more recorded music is purchased. Also, men purchase more music than women (Dilmperi, et al., 2011; Kjus, 2016).

Still, streaming services allow listeners to consume more music than before, providing them with increased options (Datta, Knox, & Bronnenberg, 2018). Consumers do not seem to use streaming services merely to sample music for future purchases, but as direct replacements for obtaining music from more traditional avenues (Wlömert & Papies, 2016). Marshall (2015) outlined three versions of streaming music services: streaming radio (curated tracks selected by the service) such as Pandora, locker services (tracks previously purchased) such as the Amazon Cloud Player, and the on-demand streaming services (user-selected tracks from a content library) such as Spotify. The increasing popularity of various streaming services required *Billboard* to retool its charting metrics to accommodate music streaming in 2012, with the RIAA and Nielsen enacting changes shortly thereafter (Andrews, 2018).

Morris and Powers (2015) note that Spotify "frames users as active agents in the process of musical discovery, even if that process involves delegating curation to the system's comprehensive data mining and algorithmic recommendation" (p. 113). This framing could find such services to help form users' self-identities, especially in social and emotional value (Oyedele & Simpson, 2018). The uses and gratifications of streaming services could help define reasons why consumers hold preferences for certain technologies over others (Im & Jung, 2016).

Active audiences are the foundation of U&G theory (Belcher & Haridakis, 2013; Katz, Blumler, & Gurevitch, 1974; Ruggiero, 2000). Uses and gratifications inform the basis of the present study, as it has been employed in previous studies investigating new media engagement, playback devices, and competition (Dunne, Lawlor, & Rowley, 2010; Ferguson, Greer, & Reardon, 2007; Gerlich, Drumheller, Babb, & De'Armond, 2015; Hollebeek, Malthouse, & Block, 2016; Whiting & Williams, 2013; Zeng, 2011). Lonsdale and North (2011) cite McQuail, Blumler, and Brown's 1972 categorization of media gratifications, which include surveillance, personal identity, personal relationships, and diversion. The present study parallels Cain's (2011) research, finding that tangible music media was still a viable option, even after the popularization of downloadable files (note: Cain's study was published prior to the popularity of music streaming services).

Methods

This study's interest specifically lies in the current uses of both physical and streaming music media, and the gratifications in which music audiences seek from these uses. Another interest is to determine if physical music media will persist. Such inquiry led to three research questions posited:

RQ1: What are streaming music subscribers currently doing with their physical music media?

RQ2: In what ways have streaming music subscribers maintained or removed their physical music media?

RQ3: Why might streaming music subscribers aim to acquire additional physical music media in the future?

After institutional review board approval, the participants for this study were initially contacted, recruited, and screened for qualification via an online participant panel service administered by Qualtrics. This service recruited U.S.-based participants that were at least 18 years and currently subscribed to a music streaming service. An online, Qualtrics-based survey with multiple-choice responses and several qualitative, open-ended questions was developed by the author and administered via the panels service in November 2018. Open-ended questions were presented to allow participants significant latitude in their descriptions (Lonsdale & North, 2011). Some of these qualitative questions—not presented to all, as it was dependent on the skip logic of the survey—included, "What did you do with your physical music purchases?," "Why did you choose to keep certain releases on (a) physical music format(s)?", and "Where do you play your physical music format(s)?" All participants were required to review the online informed consent page; if agreed, they could proceed with the survey. All data was collected anonymously.

A soft launch was first employed as a pilot to approximately ten participants to ensure proper data collection. After a successful pilot, the study was opened by Qualtrics to the qualifying population in its panels service. In total, 156 respondents submitted surveys, but some provided incomplete or unintelligible responses to the qualitative questions. After these participants' data were excluded from the review, the study was based on a population of 113 respondents who provided complete responses. Demographic information was also collected, including participants' age, gender, education, state of residence, and musical preferences.

To better ensure participants did not rush the survey, a speeding check (time minimum) was imposed, calculated from the median soft launch completion time. Participants had to answer at least 15 items (with a minimum of three qualitative responses) and—dependent on the skip logic—up to 25 items, with up to 8 qualitative responses. The average participant population age was 36 (median = 35), with range of 18–70 years. 75 participants were identified as females, 37 as males, and one as non-binary. Participants were dispersed throughout the U.S.; the most represented states included New York, Florida, Illinois, Ohio, North Carolina, Texas, and California.

Educational level was varied. Those who had completed some college were most numerous (36.3%), followed by high school educated (31%), bachelor degree (22.1%), master degree (8%), technical school (7.1%), community college graduate (7.1%), doctoral degree (1.8%), and non-disclosing (1.8%). Respondents selected their top three music genre preferences from a list of 17 popular genre options in the U.S. Hip-hop (13.1%), rock (11.9%), and country (11.2%) were the top selections, followed by classical (10%), R&B (9.1%), and blues (6.2%).

Qualitative open-ended responses were coded in two separate instances six days apart. After successive categorization rounds of qualitative responses, several themes were formed and are presented in the results.

Results

On average, participants started using subscription-based music streaming services in 2014. Pandora was used most frequently (27.4%), followed by Spotify (20.4%) and Amazon Prime (15%). 54.9% subscribed to at least two music streaming services. Participants initially learned of music streaming services from other friends (21.2%), via social media (13.3%), Google (9.7%), a music streaming app pre-loaded on their phones (9.7%), family (8.8%), or an online advertisement (8%). Other ways included radio and television advertisements, recommendations from musicians, and other websites.

Prior to subscribing to streaming music services, all participants had purchased CDs for music playback. Additionally, 17.7% purchased vinyl records and 15.9% purchased cassettes before subscribing to streaming services. A majority of the respondents (75.2%) still own their purchased music on physical formats, despite most being paying customers of music streaming services. An average of 92 albums (and/or singles) on CD, cassette, and/or vinyl were reported as currently in their collections. Separated via age groupings, the 18–29 age group had an average of 16.1 physical music recordings, the 30–39 age group had 60.2 physical music recordings on average, and the 40 and older group's average was 208.3 physical music recordings.

Most participants who played physical music media listened to them in their cars (39.8%) and on a home stereo system (30.1%). Other methods included using a computer (12.4%) and a portable player like a Walkman or Discman (4.4%). Some used a combination of devices, depending on their location. In terms of playing frequency, 22.1% of participants who owned physical music media played their CDs, vinyl, or cassettes about 2–3 times per week, while daily use was 11.5%, and 2–3 times per year at 9.7%, the third-most frequent interval. 28 respondents (24.8%) noted that they did not own purchased music on physical formats; when a follow-up question asked how they discarded their physical artifacts of music, six responded with statements such as, "I have them still" or "I kept all of them." Therefore, these six perhaps did not understand the question or may have mistakenly selected their initial response. Subtracting these 6 left 22 participants who did not own any physical music media (19.5%).

A qualitative question asked those who reported to no longer own physical music media what they had done with their recordings. This open-ended response elicited a small answer variety, with two themes of *given away* or *donations*. The most prevalent action taken with former physical music releases was that they became *donations*, oftentimes to thrift stores. Responses included, "I donated them to Goodwill," "donated to [a] thrift store," "donated to Salvation Army," and "donated to church." CDs, cassettes, and vinyl records were also *given away* to various people, including family members, as stated here: "[I] gave [them] away to my parents so they could store them at their house," "I gave them to my daughter," and "[I've] given them away to my cousins that are younger." Other responses included selling their music ("I'm pretty sure I sold some to a music store and gave the rest away to friends"), repurposing them ("I discarded them by using them for arts and crafts"), discovering that they were worn out or broken, and theft ("YouTube keeps me from buying CDs and I got robbed for all of my CDs").

Participants were also asked why they kept music on a physical format. Several themes emerged, including *nostalgia, collection maintenance, playback equipment, general enjoyment, valued items,* and *aesthetics*. Regarding *nostalgia*, participants noted that their physical music media brought them "sentimental value," it "takes me back," and because "I keep them for memories." One participant added, "some of them are scratched but they have memories I want to keep."

Others were interested in retaining a collection of physical music media, forming the *collection maintenance* theme. One said they are a collector of compact discs, another said it's a "passion or hobby," and yet another said, "I like storing them." Some participants found use in the physical music media players they kept, generating the *playback equipment* theme. Said one participant, "I still listen to my old stereo system," which was echoed by another who said, "I still own the stereo equipment and I still use it." Two more participants noted they could still play physical media in their cars.

Several positive responses indicated that participants found a *general enjoyment* in the use of their physical music media. Descriptions supporting this theme included, "I really love all the songs on the albums," "because it's good music," and that they are "albums that I completely love by artist[s] I also love." Convenience also made for an enjoyable experience. "I can listen to them while I clean house without dragging my phone around with me," said one participant, who added, "I like to listen to them as background music when I entertain."

Participants also found their physical music media to be *valued items*, whether monetarily or in terms of usability. Examples of monetary value were brought forth by this participant, who stated, "I just paid a lot for them so [I] would like to keep them longer." Another participant offered a similar take, stating they kept physical artifacts of music "since I spent money on them." Others mentioned they received physical music media as gifts, and another simply said, "I still want them."

Lastly, the physical qualities of the music media gave rise to the theme of *aesthetics*, as some listeners liked the media style, the era from which the media hailed, or its robust nature. "I like the timelessness of the physical form, and it reminds me of what we came from," said one participant, while another stated, "because I'm still old school." In terms of the lasting quality, one said, "'cause they still play good." Two more participants mentioned having a tangible product. "Love having the hard copy," said one participant. And another stated, "because I like to have art and my options."

A question asking how participants determined which physical media titles to keep received a wide range of responses. Many opted to keep the releases they generally liked ("I only buy physical copies from my absolute favorite bands") or had their favorite songs. Some noted they kept "the newest ones" or whichever recordings were most frequently listened to. One said, "I kept them all because some of them you can't always find them on streaming accounts. Some are quite old." Another kept all their physical media because "I spent money on it, and they still work, so I didn't see the need to throw them out."

While some participants had likely sought music on a streaming service that they might have already had or heard via a physical music media format, a question was posed in the opposite direction: Did playing an album or track via a streaming music service cause the participant to purchase the physical media format of that same album or track afterward? While the majority, 51.3% ($n = 58$) responded "no," 48.7% ($n = 55$) responded "yes." However, only 17 participants answered the follow-up question for the "yes" responses—to explain an instance when that occurred—in detail. Response samples included, "Listening to a favorite band's new album influenced me to purchase the vinyl," "I heard a group I liked on Pandora, so I checked them out on YouTube, loved everything, so I bought the CD"; "I streamed a song from Eminem's 'Kamikaze' and then went to Target and bought the CD"; "I streamed a Beyonce album to see if I wanted to purchase it"; and, "when Twenty One Pilots released their latest album 'Trench,' I streamed it and then ordered their vinyl album to have as a keepsake and display with my other vinyl collection."

Lastly, a question about the future of physical music format purchases revealed that a majority of participants do not anticipate making purchases of CDs, cassettes, or vinyl records in the future (57.5%). Still, a solid portion was open to the idea of future physical format purchases (42.5%). Separated by age, 54.3% of those were 18–29 years, 28.6% of those were 30–39 years, and 47.2% of those were 40 years and older. The reasons for future purchases of physical music media were varied. Four themes presented from the qualitative responses are *collectability, nostalgic inclination, tangible format preference,* and *artist preference.*

The *collectability* theme developed around participants' statements that they felt future physical music media releases are still worth keeping ("I like to keep them for a collection") and displaying ("as keepsake items to display"), and have future monetary value ("it may be worth something" and "maybe could be worth lots one day"). One participant incorporated socialization in the *collectability* of such formats: "I have friends who enjoy collecting vinyl." A lack of streaming content options was mentioned by this participant:

> As I said, I don't want to drag my phone into every room while I clean house or work in the yard. I enjoy the volume of my other devices. And mostly because you can't always get the songs I have in my collections on streaming services.

Feelings of the past helped to develop the *nostalgic inclination* theme. One participant stated simply, "[it's] nice to have for [the] memories." Another was direct about their reason for continuing to purchase compact discs: "because I'm old school and I will continue to buy my CDs in the future."

The preference of holding or having music content that is physical was another popular response, forming the theme of *tangible format preference.* A few participants stressed the media's permanency. Others were more concerned about losing access to digitally streamed media due to technological disruptions. One participant noted, "[it's] nice to have a physical copy instead of just online where it could 'glitch' and be gone." Another had a similar thought, stating, "I like to have things I like to listen to, in case I don't have Internet access."

The final theme, *artist preference,* found most participants stating that their purchases of future physical music media would be artist-dependent (or in some cases, genre-dependent). "I love to support my favorite artists," said one participant. Another was more specific: "Like if Eminem were to release, or someone I know is my favorite. If they release anything Tupac I'm buying it this time, I know that. I want to collect all of the ICP albums." And in terms of music era, this participant noted: "Sometimes that's the only way to get the classics."

Discussion

The present is an exciting period to focus on the transitional nature of music media technology, and in particular, the audience's reaction. As listeners enter a promising, burgeoning realm of streaming—with cost-effective, broad access to virtual music libraries—questions of what happens to the past (physical media) and what the future holds help to develop a more thorough understanding of music listeners' current mindsets.

In response to RQ1 (What are streaming music subscribers currently doing with their physical music media?), the majority has kept them in some quantity. Despite the recent phase-out of CDs from certain retail sales, most participants reported that a physical collection continues to persist in their possession. This also could be buttressed by vinyl usage—with sales up to 13% from the year prior (Recording Industry Association of America, 2018)—and even cassettes, with sales increasing 23% in the U.S. from 2017 to 2018 (Caulfield, 2019). Such legacy formats could be attractive to an older audience's playback experiences. Current music streaming audiences may view such tangible collections as supplemental to streaming services, especially those who noted that certain titles were solely available on physical music formats.

It could be assumed that once CD players or cassette decks malfunction, or once vehicles no longer accept CDs, such collections will contract, or perhaps disappear, due to a lack of playback mechanism. However, nostalgia could be a driving gratification, finding some resourceful audiences maintaining physical format access, especially given the median participant age of 35 (a generation likely raised on CDs). It may be another decade or two until a slimmer minority of listeners remain with digital discs in tow. Those who no longer keep physical music media stated their recordings were given away or donated—and almost always to a thrift store or charitable organization like Goodwill or the Salvation Army (see RQ2: In what ways have streaming music subscribers maintained or removed their physical music media?). This could be attributed to the ease of donations in the U.S. via routine donation collections, plus possible tax deduction incentives for donations.

Streaming subscribers were less likely to purchase more physical music media, which may be due to an overall satisfaction with their streaming experiences, especially in cases which the cost of music playback, such as on a free-to-use website like YouTube, is taken into account. Still, a sizable minority is likely to continue such purchases; they consciously (and actively) seek the gratification physical media offers (referencing RQ3: Why might streaming music subscribers aim to acquire additional physical music media in the future?). Surprisingly, more than half of participants aged 18–29 anticipated making future purchases of physical music formats, compared to just over a quarter of those in 30–39 age segment. This overall minority often labels themselves collectors, attaching themselves to the past. They may want to interact with tangible media and/or desire possession of physical music artifacts produced by their favorite artists. Such an affiliation with a physical collection may form a part of one's identity (Hall, 2007). This is also consistent with Cain's (2011) findings that those who purchased physical media were motivated to further compile their collections and reveals gratifications sought from the media they possess (Katz, Blumler, & Gurevitch, 1974). Most interestingly were those who preferred having a tangible backup; it is a prescient perspective, should they lose access to certain titles from the ever-evolving catalogs of online streaming services, or if online connections are disrupted. It is analogous to keeping a printed copy of an important digitized document.

Some of the themes appear to align with the media gratification categories by McQuail, Blumler, and Katz (1972), as used in Lonsdale and North (2011), including *nostalgia* and *nostalgic inclination* (identity and diversion), *general enjoyment* (diversion), *given away* [to family] (relationships), *collectability* (diversion), and *aesthetics* (identity and diversion). Another consistency is with Cain's (2011) findings of traditional album consumers, connecting to the theme *artist preference*, as some participants stated that purchases of future physical music media would be artist- or genre-dependent. "The respondents enjoy the work of an artist or the sound [of] a genre, so they purchase it" (Cain, 2011, p. 23). Nearly half of the overall participant population responded that a using streaming service was a motivator to purchasing a physical recording, which fits with Hiller's (2016) report of YouTube's ability to operate as a promotional vehicle for possible music sales via its recommendations.

Limitations and future directions

Due to the online data collection, the depth and verbosity of open-ended qualitative responses varied; some were diminutive phrases, while others shared several sentences. A drawback of online qualitative surveys is the lack of follow-up questioning and probing ability. Additionally, not all participants were required to respond to all qualitative questions, as a skip logic function was employed in the survey, dependent on the responses given to earlier questions. In hindsight, inclusion of further probing questions could have yielded more depth and perhaps a wider variety of perspectives, though given the administration was online, it was important to keep questioning succinct, to retain the interest of the participants.

The participant population was limited to only adults residing in the U.S. and therefore may be skewed by certain cultural and social preferences based on age or country of residence. Future studies could examine usage (or non-usage) of physical media by streaming music listeners in specific age groups and in other countries, to determine whether the results are universal or more characteristic of a particular regional preference. These studies could also cross-reference age groups and more detailed demographic factors with the decision-making processes of purchasing and usage of physical music media versus streaming music.

Note

1. All statements in this article attributable to IFPI represent the article author's interpretation of data, research opinion or viewpoints published as part of the IFPI Global Music Report 2019, and have not been reviewed by IFPI. Each IFPI publication speaks as of its original publication date (and not as of the date of this article).

References

Aguiar, J., & Waldfogel, J. (2018). As streaming reaches flood stage, does it stimulate or depress music sales? *International Journal of Industrial Organization, 57,* 278–307.

Andrews, T. M. (2018, August 12). How do we gauge success?; *Billboard*'s charts used to be our barometer. Are they meaningless in the streaming age?. *Hartford Courant*, G3.

Belcher, J. D., & Haridakis, P. (2013). The role of background characteristics, music-listening motives, and music selection on music discussion. *Communication Quarterly, 61*(4), 375–396.

Borja, K., & Dieringer, S. (2016). Streaming or stealing? The complementary features between music streaming and music piracy. *Journal of Retailing and Consumer Services, 32,* 86–95.

Cain, J. A. (2011). The album-buying niche: The future of recorded music on traditional media. *Southwestern Mass Communication Journal, 27*(1), 15–26.

Caulfield, K. (2019, January 17). U.S. cassette album sales grew 23% in 2018, aided by Britney Spears, 'Guardians,' Twenty One Pilots & more. *Billboard*. Retrieved from www.billboard.com/articles/columns/chart-beat/8493927/us-cassette-album-sales-2018-britney-spears-guardians-twenty-one-pilots

Datta, H., Knox, G., & Bronnenberg, B. J. (2018). Changing their tune: How consumers' adoption of online streaming affects music consumption and discovery. *Marketing Science, 37*(1), 5–21.

Dilmperi, A., King, T., & Dennis, C. (2011). Pirates of the web: The curse of illegal downloading. *Journal of Retailing and Consumer Services, 18,* 132–140.

Dunne, A., Lawlor, M. A., & Rowley, J. (2010). Young people's use of online social networking sites — a uses and gratifications perspective. *Journal of Research in Interactive Marketing, 4*(1), 46–58.

Edmond, M. (2014). Here we go again: Music videos after YouTube. *Television & New Media, 15*(4), 305–320.

Ferguson, D. A., Greer, C. F., & Reardon, M. E. (2007). Uses and gratifications of MP3 players by college students: Are iPods more popular than radio? *Journal of Radio Studies, 14*(2), 102–121.

Ford-Stewart, J. (2018, March 1). Increasingly CD players in cars are disappearing just like tape decks once did. *Journal Sentinel.* Retrieved from www.jsonline.com/story/communities/southwest/news/2018/03/01/increasingly-cd-players-cars-disappearing-just-like-tape-decks-once-did/355861002/

Gerlich, R. N., Drumheller, K., Babb, J., & De'Armond, D. (2015). App consumption: An exploratory analysis of the uses & gratifications of mobile apps. *Academy of Marketing Studies Journal, 19*(1), 69–79.

Given, J. (2015). Owning and renting: Speculations about the past, present and future acquisition of audiovisual content by consumers. *Studies in Australasian Cinema, 9*(1), 21–38.

Hall, A. (2007). The social implications of enjoyment of different types of music, movies, and television programming. *Western Journal of Communication, 71*(4), 259–271.

Herbert, D., Lotz, A. D., & Marshall, L. (2019). Approaching media industries comparatively: A case study of streaming. *International Journal of Cultural Studies, 22*(3), 349–366.

Hiller, R. S. (2016). Sales displacement and streaming music: Evidence from YouTube. *Information Economics and Policy, 34,* 16–26.

Hollebeek, L. D., Malthouse, E. C., & Block, M. P. (2016). Sounds of music: Exploring consumers' musical engagement. *Journal of Consumer Marketing, 33*(6), 417–427.

IFPI. (2019, April 2). IFPI Global Music Report 2019. International Federation of the Phonographic Industry. Retrieved from www.ifpi.org/news/IFPI-GLOBAL-MUSIC-REPORT-2019

Im, H., & Jung, J. (2016). Impacts of personal characteristics on the choice of music consumption mode: purchasing CD, downloading, streaming, and piracy. *Journal of Media Business Studies, 13*(4), 222–240.

Ironmonger, D. S., Lloyd-Smith, C. W., & Soupourmas, F. (2000). New products of the 1980s and 1990s: The diffusion of household technology in the decade 1985–1995. *Prometheus, 18*(4), 403–415.

Katz, E., Blumler, J. G., & Gurevitch, M. (1974). Uses and gratifications research. *Public Opinion Quarterly, 37,* 509–523.

Kelley, C. (2019, April 29). Spotify reaches 100 million paid subscribers ahead of Apple Music. *Forbes.* Retrieved from www.forbes.com/sites/caitlinkelley/2019/04/29/spotify-100-million-paid-subscribers-apple-music/#623f3e5d117f

Kjus, Y. (2016). Reclaiming the music: The power of local and physical music distribution in the age of global online services. *New Media & Society, 18*(9), 2116–2132.

Knopper, S. (2018, June 14). The end of owning music: How CDs and downloads died. *Rolling Stone.* Retrieved from www.rollingstone.com/music/music-news/the-end-of-owning-music-how-cds-and-downloads-died-628660/

Kot, G. (2003, March 30). Twilight for the CD. *Chicago Tribune.* Retrieved from www.chicagotribune.com/news/ct-xpm-2003-03-30-0303290329-story.html

Kreps, D. (2019, April 8). Jeff Tweedy to release 'Warm' companion LP 'Warmer' on Record Store Day. *Rolling Stone.* Retrieved from www.rollingstone.com/music/music-news/jeff-tweedy-warmer-record-store-day-family-ghost-819004/

Kumar, K. (2018, July 25). Best Buy faces the music with CDs. *Chicago Tribune,* 2.

Liikkanen, L. A., & Salovaara, A. (2015). Music on YouTube: User engagement with traditional, user-appropriated and derivative videos. *Computers in Human Behavior, 50,* 108–124.

Lonsdale, A. J., & North, A. C. (2011). Why do we listen to music? A uses and gratifications analysis. *British Journal of Psychology, 102*(1), 108–134.

Marshall, L. (2015). 'Let's keep music special. F—Spotify': On-demand streaming and the controversy over artist royalties. *Creative Industries Journal, 8*(2), 177–189.

McIntyre, H. (2017, March 2). It's time to say goodbye to the CD player in new American cars. *Forbes.* Retrieved from www.forbes.com/sites/hughmcintyre/2017/03/02/the-time-has-come-to-say-goodbye-to-the-cd-player-in-new-american-cars/#51bf220c2e88

McQuail, D., Blumler, J. G., & Brown, J. R. (1972). The television audience: A revised perspective. In D. McQuail (Ed.), *Sociology of mass communications* (pp. 135–165). Harmondsworth: Penguin.

Meier, L. M., & Manzerolle, V. R. (2019). Rising tides? Data capture, platform accumulation, and new monopolies in the digital economy. *New Media & Society, 21*(3), 543–561.

Mervis, S. (2018, February 23). Compact disc(ontinued): As sales decline and stores stop carrying them, should you hang on to your CDs? *Pittsburgh Post-Gazette,* A-1.

Miers, J. (2011, December 4). Death by download: As some predict the death of the CD, streaming sites like Spotify and rDio raise the question of whether music one day will be entirely digital. *The Buffalo News,* F1.

Mitchell, D. M., Scott, C. P., & Brown, K. H. (2018). Did the RIAA's prosecution of music piracy impact music sales? *Atlantic Economic Journal, 46,* 59–71.

Morris, J. W., & Powers, D. (2015). Control, curation and musical experience in streaming music services. *Creative Industries Journal, 8*(2), 106–122.

Oyedele, A., & Simpson, P. M. (2018). Streaming apps: What consumers value. *Journal of Retailing and Consumer Services, 41,* 296–304.

Palladino, V. (2019, April 5). More people pay for Apple Music than Spotify in the US now. *Ars Technica.* Retrieved from https://arstechnica.com/gadgets/2019/04/report-apple-music-surpasses-spotify-with-28-million-paid-us-subscribers/

Poort, J., & Weda, J. (2015). Elvis is returning to the building: Understanding a decline in unauthorized file sharing. *Journal of Media Economics, 28*(2), 63–83.

Recording Industry Association of America. (2018). *Mid-year 2018 RIAA Music Revenues Report.* Retrieved from www.riaa.com/wp-content/uploads/2018/09/RIAA-Mid-Year-2018-Revenue-Report-News-Notes.pdf

Roberts, R. (2018, July 26). A new tune for YouTube; its foray into music streaming carries with it a set of advantages and challenges. *Los Angeles Times,* E-2.

Rondan-Cataluna, F. J., & Martin-Ruiz, D. (2010). Customers' perceptions about concerts and CDs. *Management Decision, 48*(9), 1410–1421.

Ruggiero, T. E. (2000). Uses and gratifications theory in the 21st century. *Mass Communication & Society, 3*(1), 3–37.

Shaw, L. (2019, May 8). Google tops 15 million music subscribers as it chases Spotify. *Bloomberg.* Retrieved from www.bloomberg.com/news/articles/2019-05-08/google-tops-15-million-music-subscribers-as-it-chases-spotify

Sherburne, P. (2016, July 13). The holdouts: An exploration of vinyl-only labels in the digital age. *Pitchfork.* Retrieved from https://pitchfork.com/features/article/9915-the-holdouts-an-exploration-of-vinyl-only-labels-in-the-digital-age/

Sinclair, G., & Tinson, J. (2017). Psychological ownership and music streaming consumption. *Journal of Business Research, 71,* 1–9.

Sisario, B. (2015, June 12). Music streaming service aims at Japan, where CD is still king. *New York Times,* B2.

Sodomsky, S. (2018, July 2). Jack White announces new vinyl-only live album. *Pitchfork.* Retrieved from https://pitchfork.com/news/jack-white-announces-new-vinyl-only-live-album/

Steele, A. (2017, November 8). In the streaming age, Taylor Swift plugs 'Reputation' on CD; The pop star's new album is unlikely to be available to stream when it comes out on Friday. *Wall Street Journal* (Online). Retrieved from www.wsj.com/articles/in-the-streaming-age-taylor-swift-plugs-reputation-on-cd-1510182069

Steele, A. (2018, July 26). Spotify reports increase in subscribers amid promotions; music-streaming company posts wider loss, tempers subscriber-growth expectations. *Wall Street Journal (Online).* Retrieved from www.wsj.com/articles/spotify-reports-increase-in-subscribers-amid-promotions-1532604544

Stutz, C. (2017, November 30). Taylor Swift's 'Reputation' finally headed for streaming services on Friday: Sources. *Billboard*. Retrieved from www.billboard.com/articles/columns/pop/8054705/taylor-swift-reputation-streaming-services-friday

Swanson, K. (2013). A case study on Spotify: Exploring perceptions of the music streaming service. *MEIEA Journal, 13*(1), 207–230.

Thomes, T. P. (2013). An economic analysis of online streaming music services. *Information Economics and Policy, 25*(2), 81–91.

Thrasher, D. (2012, January 15). Digital downloads top CD sales for first time. *Dayton Daily News*, D-18.

Tilley, S. (2018, August 3). Streaming — has it killed music as we know it? *Florida Today*, A4.

Waldfogel, J. (2010). Music file sharing and sales displacement in the iTunes era. *Information Economics and Policy, 22*, 306–314.

Watson, A. (2019a, August 9). Share of music streaming subscribers worldwide as of the first half of 2018, by company. *Statista*. Retrieved from www.statista.com/statistics/653926/music-streaming-service-subscriber-share/

Watson, A. (2019b, August 9). Number of Deezer's paying subscribers worldwide as of January 2019. *Statista*. Retrieved from www.statista.com/statistics/321559/deezer-paying-subscribers/

Whiting, A., & Williams, D. (2013). Why people use social media: A uses and gratifications approach. *Qualitative Market Research: An International Journal, 16*(4), 362–369.

Wlömert, N., & Papies, D. (2016). On-demand streaming services and music industry revenues — insights from Spotify's market entry. *International Journal of Research in Marketing, 33*, 314–327.

Zeng, L. (2011). More than audio on the go: Uses and gratifications of MP3 players. *Communication Research Reports, 28*(1), 97–108.

Index

Note: **Bold** page numbers refer to tables; *italic* page numbers refer to figures and page numbers followed by "n" denote endnotes.

A/B testing 44
active audiences 116
Actor-Network Theory 52
aesthetics theme 119, 121
affordances 66–67
Airoldi, A. 38
album releases 72
algorithm-curated playlists 86, 88
algorithmic affordances 67, 75–76
algorithmic culture 100, 102–103, 110
algorithmic individuation 80
algorithms 43, 66–67, 75, 84, 85; kinds of 88; metrics and 56; music streaming 38–41; newsgathering 85; in social media 85; for song performance 70; streaming 65, 67; studies of 103
Allington, D. 31n12
Amazon 2
Amazon Prime 118
Anderson, Chris 2, 83
Anghami 25
Apple Music 36, 41, 44, 56, 114; playlist curator 54
Application Programming Interface (API) 15n4
apps: "Every Noise at Once" 10; Forgotify 8, 8–10, 9; infomediary 3; innovative 3; music 2; "Music Popcorn" 2, 3, 9–10, 10; Spotify-linked (companion) 3
artist preference theme 120
artists: *Brazilian Funk* 89–91, 89–91; independent 86; male *vs.* female 110n3; *Sertanejo* 91–93, 92–95
Artists and Repertoire (A&R) 24
audience action 84
audience approval 84
Audience-Media Engine 84, 85, 94
audience reach 84
audio-only format music 4
Australian music industries 49, 50, 58; context and interview sample 49–51; locating in-house curated Spotify playlist 51–52
Australian Research Council 49
Avdeeff, M. 108

Bandcamp 10
Baym, Nancy 59, 80
B2B interfaces 65
Beirut, independent music 19–29
Beirut Jam Sessions 23
Beraldo, D. 38
Big Data systems 84
Billboard 56
Bishop, S. 104
Bivens, R. 104
black box 38–41
Block, M. P. 114
Bloomberg Businessweek 54
Blu Fiefer 22
Blumler, J. G. 121
Bob Marley 114
Borja, K. 116
Bourdieu, Pierre 51, 52
boyd, D. M. 21
Braidotti, R. 100, 101
Brayne, S. 103
Brazilian Funk 82, 88; artists 89–91, 89–91
Brazilian music industry 81–82; *Brazilian Funk* artists 89–91, 89–91; digital music gatekeeping 85–88, 87; gatekeeping, datacracy, and platformization 83–85; genres 95; methodology 88–89; *Sertanejo* artists 91–93, 92–95; understanding *Brazilian Funk* and *Sertanejo* 82
Brian (pseudonym) 56, 57
broad folksonomy 7
Brown, James 82
Brown, J. R. 121
browse function 101, 108–109
Bruno Mars 44
Bucher, T. 39, 41, 44
Burkart, P. 79, 80, 86
business-intelligence resources 69
Byrne, David 40, 53

Cain, J. A. 116, 121
Cairo, A. 5
Carter, Troy 43
Celestial Jukebox 79
centralized agents 88
centralized gatekeepers 86
Chapman, Tracy 106, 109
charts 13
Chery, Carl 54
Chun, W. H. K. 103
Chyno 32n14
cloud-based music streaming 4
"cold start problems" 71
collaborative filtering 8, 14, 15n2
collectability theme 120
collection maintenance theme 119
Collins, P. H. 101
communication processes 79
compact discs (CDs) 113, 114, 118; Japanese
    consumers 116; purchases of 120; sales of 115;
    see also physical music media
content, information ecology 3
context, information ecology 3
Cooke, C. 40, 41
copyrights 40
Couldry, N. 66, 75
crowd-funding platforms 19
Crowdy, D. 50
Cukier, K. 66
cultural capital 51, 55
cultural intermediaries 30n1, 51–52
Culture Unbound (Fleischer and Snickars) 24
"culture workers" 30n1
Cuming, Lou 53
curated playlist 52–53
cyclical/processual datafication function 75

Daboh, Austin 41
data: competitive advantage of 71;
    increasingly detailed 68–71; innovations
    in presentation of 3; interpretations of 70;
    interview 52, 54, 55; and novelty 73–75
data colonialism 83–84
datacracy 80, 81, 83–85, 93
datacratic-oriented regime 84
datacratic regime 80
data-driven customization of culture 80
datafication 65–67, 69, 75–76, 83
Deady, Beth 55, 57
decentralized agents 88
decentralized gatekeepers 86–87
Deezer 114
Del.icio.us 2
Demand-Side Platforms 60n4
Dennis, C. 116
"describing tools" 2
Dieringer, S. 116

digital gatekeeping model 85, 95
digital gatekeeping theory 94
digital infomediaries 52
digital journalism 84–85
digital media 103
digital music gatekeeping 85–88, 87
digital music industry 72
digital revenues 36
digital service providers (DSPs) 40, 51–55, 59, 60
digital technology 35, 67, 81
Dilmperi, A. 116
discover function 101, 107
Discover Weekly 85, 88, 93
discovery 51–54, 57, 59, 60n6
Ditto music 55
DJ Khaled 96n7
DJ Semtex 41
Drake, Nick 106
drop-down menus 5
DSPs see digital service providers (DSPs)
Dueck, B. 31n12
Durham, B. 24
Dylan, Bob 106

The Echo Nest 2, 3, 9–10, 13, 14, 71
ecosystem evolution 79
EDM see electronic dance music (EDM)
Egypt 31n11; SoundCloud in 27–28
Ek, Daniel 9, 40, 43, 44, 106
electronic dance music (EDM) 4, 10, 108
electronic music area 10
ElectroZajal event 31n5
Ellison, N. B. 21
El Rass 32n14
engagement: lack of artist/fan 58, 60; playlists and
    57–59; with social theory 49
enthusiasm 53–54
"Envolvimento" 89
Ergatoudis, George 41
Eriksson, M. 101
European Commission 58
Evans, M. 50
event-driven music distribution 72–73
events 71–73
"Every Noise at Once" 10

Facebook 21–22, 28, 44, 85, 87, 104;
    algorithms 66
"fake artists" 74
featured content 6
Federal organizations 50
filter bubbles 83, 85
Fincher, David 37
Fleischer, R. 15n11; Culture Unbound 24
Flickr 2
folksonomous tag clouds 4
folksonomy 1, 2, 4; elements of 14; Forgotify and 9

folksonomy-friendliness 6; features 14; functions 1; sites 7
folksonomy-friendly wayfinding features 4–5; drop-down menus 5; featured content 6; headers 5; player bar 6–8, 7; queue 6; search filters 6; tag clouds 5; text-box search 5–6
Forgotify 8, 8–10, 9, 13, 15n3
Fournet, A. 31n8
The Frames 106
"Fresh Finds" 8
Frew, Ronnie 57
front-facing user interfaces, of music-streaming services 66
fundamental information asymmetry 39
funding: Australian music industries 50; of independent music 20
Furacão 2000 82

Gaffney, M. 2
Gandini, A 38
Garland, S. 31n8
gatekeeper agents 85
gatekeepers: centralized 86; decentralized 86–87; of music 83; types of 85
gatekeeping 80, 83–85, 88
gatekeeping theory 83
gender: identities in systems of 103; material and symbolic dimensions of 101; organizes social and cultural life 101; software effects in social media 104; technology and 101
gender binary 104
gendered music streaming 109
general enjoyment theme 119, 121
genre maps 2
genre rules 106
genres 2, 10, 11, 52, 81, 82, 88; Brazilian music genres 95; gendering and racialization of 109; K-pop 74; self-reported artist 60n2; traditional 38
geotags 2
Gibbins, N. 52
Gibson, J. J. 67
Gillespie, T. 39, 67
Given, J. 114
Glitter, Gary 43, 44
Global Music Report (2019) 113
GoldLink 28
Gonzales, José 106
Goodwill 121
Google News 85, 88
Gray, David 106
Guns N' Roses 114
Gutenberg, Johannes 79
Gutenberg parenthesis 79, 85, 86

Hagen, A. N. 11
Halford, S. 52
Hamdan, Zeid 18, 29, 30

Hamilton, Sarah 55, 57
Hannigan, Lisa 106
Hansard, Glen 106, 107
headers 5
Hegarty, Antony 105
Hepp, A. 66, 75
Hesmondhalgh, D. 20
Hiller, R. S. 122
Hirsch, Paul M. 83, 86, 93
Hogan, B. 66–67, 75
Hollebeek, L. D. 114
Hope in the Dark (Solnit) 12, 12
Horton, S. 7
House of Cards 36–37, 39
Hracs, B. J. 52
Hu, Cherie 13, 59
Hughes, D. 50
Hugo & Guilherme 92, 93
human-computer interaction 3
human goals 3
humans vs. automated algorithmic systems 55

IA see Information Architecture (IA)
independent artists 55, 86, 90
independent music: Beirut 19–29; social media for 21–24
independent musicians 19, 31n5
individual amateurs 85, 87
industry-related playlists 72
infomediary 52
infomediary apps 3
informants 69, 71–72, 74; data and distribution practices 68; digitization's influence on professional practice 70; recruitment of 67; regular reliance on real-time data 73; Spotify 75; use metrics 75
Information Architecture (IA) 3
information ecology 3
Ingham, T. 81
in-house curated playlists 52–53, 59, 93
in-house curated Spotify playlist, locating of 51–52
innovations 3
innovative apps 3
"in real life" ("IRL") 19, 23
Instagram 22, 87
interfaces 1; B2B, 65; and companion apps 8–14; front-facing user 66; invisible 64; web 4
interlocutors 20, 22, 23, 30n2
intermediaries: cultural 30n1, 51–52; stakeholders and 68
intermediation 52, 55
International Federation of the Phonographic Industry (IFPI) 36, 113
Internet 21, 78
interpretations 68; of data 70
interpretive knowledge 68

interview data 52, 54, 55
invisibility 41–45
invisible interface 64
Irglova, Marketa 106

Jakovljevic, Jana 12
Johansson, A. 101
Jordan, Lane 8
Jordanous, A. 31n12
journalists 85

Keith, S. 50
Kelly, R. 42–44
Khalifeh, Marcel 31n4
Khuluki, Tarek 22
King, T. 116
KondZilla 82, 89–92
K-pop genre 74, 107
Kymäläinen, T. 3

labelling 4; in music streaming 4
Lamar, Kendrick 43
language ideology 31n11
LANKS 53, 54
The Laws of Simplicity (Maeda) 3
Lebanon, streaming platforms in 24–26
less-experienced performers 53
Lewin, Kurt 83
literature review 114–116
Long tail 2, 83
Lonsdale, A. J. 114, 116
Lostprophets 43, 44
Lowe, Zane 41
Lucca & Mateus 92, *92, 94*
Luvaas, B. 31n8
Lynch, P. J. 7

MacDonald, Glenn 9, 10, 15n8
Mackenzie, A. 103
McNeil, M. 101
McNicol, A. 104
McQuail, D. 121
Maeda, J.: *The Laws of Simplicity* 3
Malthouse, E. C. 114
Manovich, L. 4
Marshall, L. 54, 116
mass media 79
Maund, Chris 58
Mayer-Schönberger, V. 66
MC Dede 91, *91*
MC Loma & as Gêmeas da Lacração 89, *89, 90*
media presence 84
Meier, L. M. 20
metrics: economic benefits of 69; informants use 75; real-time 73
MGM Distribution 55

monetization of music 60n1
Moore, C. 102
"Morning commute" 108
Morris, J. W. 1, 2, 83, 102, 116
Morrow, G. 50
Morville, P. 3
MSSs *see* music-streaming services (MSSs)
Mulligan, M. 40, 59
multi-node model 85
Mushroom Group 58
music: gatekeeper of 83; offered via digital platforms 79; online distribution of 65
music apps 2
music-based social media apps 2–3
music business environment, transformation of 79
music catalogs 79
music discovery 1–4, 8, 9, 14, 15, 37, 54
"music for relaxation" 38
Music Genome Project 24
music genres 108, 110; folksonomy-born vocabularies of 4; immensity of 11
musicians: in Beirut 22–23; marketing of 24; Romantic perspective on 38
music industries 60n1, 64
music industry 35, 78, 80; focus on novelty in 74–75; influenced by market demands 84; recurring events in 72
music listening 100; software for 100–101
music metrics 69
"Music Popcorn" 2, 3, 9–10, *10*
music purchase habits 115
music realm 84
music sales data 80
music streaming interfaces: algorithms 38–41; growth of 35–38; labeling in 4; wayfinding features in 3–4; wayfinding features within 7
music-streaming services (MSSs) 37, 49, 64, 100; competitive advantage of data 71; events 71–73; fresh data and novelty 73–75; front-facing user interfaces of 66; impact on user' behavior 65; increasingly detailed data 68–71; methods 67–68; reinforcing feedback loops and algorithmic affordances 75–76
music streaming sites 14
#MuteRKelly social media campaign 42

Nakamura, L. 103
Napoli, Philip 79
Napster 36
narrow folksonomy 7
Negus, K. 51, 79, 83, 86
Netflix 36–37, 44
"New music Friday" 108
newsgathering algorithms 85, 88
Nieborg, D. B. 79, 80, 85
Nielsen Music report 115

non-disclosure agreements 40
non-economic capital 51
North, A. C. 114, 116
Norwegian music business 65–66
Norwegian music industry 67
nostalgia theme 119, 121
nostalgic inclination theme 120

on-demand platforms 64
O'Neil, C. 39
Onfroy, Jahseh Dwayne 42
online data collection 122
online music 65, 102; studies of 114
open-ended questions 117
organizational systems 1
Otavio Augusto & Gabriel 93, *95*
outright ban 42

paid-curated content 6
Pandora 24, 116, 118, 119
Pandora Internet Radio 38
panopticon 41–45
Pariser, Eli 85
Pasick, Adam 8
Pasquale, F. 38
passive listening 58
Paul Lamere 9, 10, 15n5
peer-to-peer (P2P) file sharing networks 116
Pelly, Liz 56
Perry, Katy 44
pessimism 53–54
Pettitt, T. 79
Phil (pseudonym) 58
phonograph effect 80
physical music media 118; collection of 119; sales
    decline of 115–116; *see also* compact discs (CDs)
Pink Harvest 57
pitching process 59, 60n7
platform 60n3
"platform capitalism" 39
platformization 83–85
Playax 80, 81, 88, 89, *89*, *91–93*, 93, *95*, 96n10
Playax Ranking 88–89
playback equipment theme 119
player bar 6–8, *7*
playlist events 72
playlist pitching 51, 55–57
playlists 11–14, 79, 83, 92, 104, 108; access to
    promotion 54–57; algorithm-curated 86;
    curated 52–53; and engagement 57–59;
    industry-related 72; revenue on 53–54
playola, accusations of 56
Poell, T. 79, 80, 85
"Popcorn Music" 10
Postcards 25, 26, 30
Postman, Neils: *Technopoly* 83

"post-playlist reality" 59
Powell Books 12, *12*
Powers, D. 1, 2, 83, 116
Prey, R. 37, 80, 83
process knowledge 68
public downgrade 42, 44
public upgrade 42

qualtrics-based survey 117
queue 6

race 101, 103–110
Rafferty, P. 2
Rahbani, Ziad 31n4
Ray, Mary Beth 57
real-time metrics 73–74
recorded-music revenues 64
recording industry 48–49; stardom, risk-reducing
    economic structures 50
Recording Industry Association of America
    (RIAA) 115, 116
reinforcing feedback loops 75–76
related artists function 101, 104–107
Release Radar 15n1, 72
Reputation 115–116
revenue: digital 36; global music subscription
    audio streaming 113; on playlists 53–54
revenue-generating streaming services 51
Rice, Damien 100, 102, 105–107, 109
Rogers, J. 35
Rose, T. 109
Rosenfeld, L. 3

Sabra, Julia 25
Salvation Army 121
Sauerberg, L. 79, 85
screenshots 101, 102
scrollable content 6
search filters 6
Seaver, N. 66, 84
secondary orality 79
self-reported artist genres 60n2
Semaan, Anthony 23
sensationalist rhetoric 48
Sertanejo 82, 88, 89; artists 91–93, *92–95*
Sertanejo Universitário 91
shadow ban 42, 44
shadow downgrade 42, 44
shadow upgrade 42
Shaw, Lucas 54
Sheeran, Ed 44
Shoemaker, P. 84
side-bar menus 5
Sloth, Charlie 41
Snickars, P.: *Culture Unbound* 24
social-based music sharing sites 2–3

social capital 51, 55
social media 78, 87–88; algorithms in 85; feminist
     studies of software and 103–104; rising of
     popularity 84
social media platforms 19, 28, 29; for
     independent music 21–24; music fans in 79;
     tracking of users 80
social-streaming hybrid 26–29
sociotechnical cultural intermediary 51–52
software: for music listening 100–101; for music
     streaming investigation 102–103; and social
     media, feminist studies of 103–104
Solnit, Rebecca 15n10; *Hope in the Dark* 12, *12*
Songza 3
SoundCloud 10, 11, 18–19, 26–30, 31n12, 32n13,
     36, 87, 88
"SoundCloud community" 28
SoundScan 80
Spacey, Kevin 37
splits 41
Spotify 2, 3, 19, 24, 30, 36, 38, 40–44, 55, 57,
     60n3, 66, 75, 80–81, 86, 88, 92, 93, 100,
     102, 114, 116, 118; discover page *13*, 13–14;
     in-house curated playlist 49; interface and
     companion apps 8–14; mobile and desktop app
     11–12; New Music Friday 72; Release Radar
     72; user interface (UI) 1; wayfinding features
     in 11, **11**, *11*
Spotify analytics 69, 74
Spotify curated playlists 90–91
Spotify for Artists portal 59, 69, 71, 73, 84
Spotify functions: browse 101, 108–109; discover
     101, 107; related artists 101, 104–107
Spotify-linked (companion) apps 3
Spotify playlists 52, 56, 59
Spotontrack 74
Sprengel, D. 31n11
Srnicek, N. 39
Stack, L. 38
stakeholders 65; in Australian music industries
     49; cultural and social capital of 55; as
     expert interviewees 68; and intermediaries
     68; at music companies 51, 75; Spotify
     relations with 71; streaming technology
     supplies 65
stardom, risk-reducing economic structures 50
strategic professionals 85, 87, 95
Stratton, J. 105
streaming algorithms 65, 67
streaming music services, growth of 113
streaming platforms 35, 83; in Lebanon 24–26
streaming playlists 58
streaming subscribers 121
Striphas, T. 100, 103, 106
subscription-based music streaming services 118

surveillance capitalism 41
Swanson, K. 115
Swift, Taylor 40, 53, 86, 115–116

tag clouds 4–6
tangible format preference theme 120
Tanjaret Daghet 22
Tarantino, Quentin 9
Target.com 114
Target's Black Friday sale 114
"taste influence" 78
taxonomy 1, 2
technology: digital technology 35, 67, 81; and
     gender 101; progression of 115; streaming 49,
     65; visualization 5
Technopoly (Postman) 83
text-based tags 2
text-box searches 5–6; Spotify 13–14
third-party playlists *see* in-house curated playlists
Tidal 115
time tags 2
Tohme, Marwan 25
traditional genres 38
Triple J 50
Trump, Donald 42
Tweedy, Jeff 116
Twitter 87
two-faced virtual ecosystem 19

U&G theory 116
"unbundling" albums 36
United States, digital revenues in 36
unlistened music 14
uploading process 31n10
user interface (UI) 1, 37
user-made metadata 15
users: behavior, impact of MSSs 65; of Forgotify
     9; functions for 13; information ecology 3;
     social platforms tracking of 80; of Spotify 109
UX design 3; discourse of 4
UX tool, of wireframes 1, 5

valued items theme 119
Vander Wal, Thomas 2, 7
Van Dijck, J. 66
vaporwave 10, *10*
vinyl LPs 113, 114
VKontakte 102
Vonderau, P. 60n4
Vos, T. 84

Wajcman, J. 101
Wallace, J. 84–86, 95
Warmer 116
wasta 20

wayfinding features 3–4
wayfinding functions 7
Web 2.0 2, 21, 104
web interfaces 4
websites 1, 22, 25, 26, 114, 121
Webster, J. 52
White, David M. 83
White, Jack 116
WhoSampled 3
Wickerpark 23
Wikström, P. 35, 84, 85, 94

Winehouse, Amy 114
wireframes 5; UX tool of 1, 5
"The Wreckomender" 3

XXXTentacion 42, 43

Yorke, Thom 40, 53
YouTube 36, 86, 89–93, 95, 96n6, 102, 104, 108,
    113, 121, 122
YouTube for Artists 84
YouTube Music 114